Praise for

# *Gulp*

Amazon's Best Science Book of 2013
A *Publishers Weekly* Best Book of 2013
A *Booklist* Top Ten Book of the Year
A *Washington Post* Notable Book of 2013
A Goodreads Choice Awards Best Nonfiction Book of 2013
An iBooks Best Book of 2013
A Bookpage.com Top 50 Book of 2013
A Gizmodo.com Best Book of 2013
A *Seattle Times* Best Title of 2013
A *Globe & Mail* Best International Non-Fiction
Book of the Year
A *New Statesman* Best Title of 2013
A BrainPickings.org Best Science and Technology
Book of 2013
A Bookrageous.com 2013 Favorite

"Far and away her funniest and most sparkling book, bringing Ms. Roach's love of weird science to material that could not have more everyday relevance. . . . Never has Ms. Roach's affinity for the comedic and bizarre been put to better use. . . . *Gulp* is structured as a vastly entertaining pilgrimage down the digestive tract, with Ms. Roach as the wittiest, most valuable tour guide imaginable." —Janet Maslin, *New York Times*

"There is much to enjoy about Mary Roach—her infectious awe for quirky science and its nerdy adherents, her one-liners. . . . She is beloved, and justifiably so." —Jon Ronson, *New York Times Book Review*

"Like the perfect dinner guest full of entertaining conversation— or wait, given the subject, let's delay this until dinner is over— Roach rolls out one surprising story after another. . . . Part of the fun of reading Roach is watching her teeter wildly along the borders of tastelessness." —*Chicago Tribune*

"An enjoyable romp. . . . [Roach] has a keen eye for the fascinating and often gruesome anecdote." —*The New Yorker*

"Ms. Roach can hardly contain her enthusiasm for the details of digestion. It's not difficult to imagine her in 19th-century explorer's garb, standing on the prow of a tiny boat sailing from the tip of the tongue to the unknown and perilous reaches beyond. . . . The result is a tale that can be revolting but that inexorably draws the reader along with peristaltic waves of history and vividly described science." —*Wall Street Journal*

"I inhaled this terrific, offbeat book in a couple of sittings, and I've been gleefully regurgitating its fascinating insights and astonishing, delightfully repulsive anecdotes ever since. . . . Roach's enthusiasm and wit are infectious." —*San Francisco Chronicle*

"Gulp is a whirlwind tour. Luckily, Roach makes you feel as if you're on the tour bus with your funniest friend providing running color commentary. She's completely fearless, perfectly happy

to plunge her hand—nay, her entire arm—into a cow's stomach
to feel its digestive caress."                    —*Science News*

"Roach guides the reader gently down the upper reaches of the
alimentary canal like a trustworthy gondolier before hitting the
bawdy, chaotic rapids nearer the end of the journey. . . . So much
nauseating fun."                                  —*Slate*

"Brilliantly mischievous."                        —*Bloomberg*

"Roach brings a good-natured and (I dare say) feminine empathy
absent from most science writing. . . . [She] takes us inside her
sensibility (part Katharine Hepburn, part Nora Ephron) and
binds us to her emotionally. . . . The country's greatest popular
science writer."                        —*Los Angeles Review of Books*

"Mary Roach's closest analogues in the bookstores are really the
uber-intelligent sometime-humorists more often shelved in the
essay section, near the poetry and the fiction, her true kind: David
Foster Wallace, Ian Frazier, Lawrence Weschler, Steve Almond."
                                                  —*Salon*

"Gulp is about revelling in the extraordinary complexities and
magnificence of human digestion."                —*Economist*

"Relentlessly fun to read."                       —*New Republic*

"[Roach] is a good journalist: hard-working, endlessly curious,
irreverent, attentive, sceptical. She illuminates a tricky but hugely
fascinating subject while rarely blundering across the borders of

reasonable taste. Her book is—dare I say it?—the best kind of lavatory reading." —*Telegraph*

"A touchy topic illuminated with wit and rigor, packed with all the stinky details." —*Kirkus Reviews*

"Never before has the process of eating been so very interesting. . . . After digesting her book, you can't help but think about what that really means." —*Pittsburgh Post-Gazette*

"One of my top criteria for pronouncing a book worthwhile is the number of times you snort helplessly with laughter and say, 'Wow! Did you know that . . .' before your long-suffering spouse throws a book at you from across the room. My personal spouse says that, in this department, *Gulp* takes the cake."—*Seattle Times*

"Filled with witty asides, humorous anecdotes, and bizarre facts, this book will entertain readers, challenge their cultural taboos, and simultaneously teach them new lessons in digestive biology." —*Library Journal*, starred review

"Stomach-churning at times and thoroughly enlightening at others, an expedition devoted to the overlooked milieu of a topic considered foul." —*Minnesota Daily*

"It's as gross as one might expect. But it's also enthralling. . . . It's clear she revels in giving readers a thrill—even if it is a queasy one." —*Publishers Weekly*, starred review

"By turns, it fascinates, grosses out . . . and thoroughly entertains. I'm not sure whether I grimaced or guffawed more, but I'm still surprised no fellow BART passengers who shared my train as I read *Gulp* asked what on earth provoked such reactions from me. . . . Like the Olympic gymnast whose incredible control makes a complex maneuver seem effortless, Roach conceals such painstaking research and paraphrase with her witty style and often-irreverent footnotes."                                    —*Paste*

"Few writers are able to write about science in a way that's provocative without being sensationalistic, truthful without being dry, enchanting without being forced—and even fewer are able to do so on subjects that don't exactly lend themselves to Saganesque whimsy. After all, it's infinitely easier to inspire awe while discussing the bombastic magnificence of the cosmos than, say, the function of bodily fluids and the structures that secrete them. But Mary Roach is one of those rare writers, and that's precisely what she proves once more in *Gulp* . . . a fascinating tour of the body's most private hydraulics."        —Maria Popova, Brainpickings.org

# Gulp

· ADVENTURES ON THE ALIMENTARY CANAL ·

# Mary Roach

W. W. NORTON & COMPANY

New York ∗ London

Photograph Credits: Introduction: © SuperStock / SuperStock / Getty Images; Chapter 1: © Bettmann/Corbis; Chapter 2: © Bettmann/Corbis; Chapter 3: © ClassicStock/Corbis; Chapter 4: © Paul Garnier / fStop / Getty Images; Chapter 5: © Bettmann/Corbis; Chapter 6: © CSA Images / Archive / CSA Images / Getty Images; Chapter 7: © Ralph Crane / Time & Life Pictures / Getty Images; Chapter 8: © Fox Photos / Hulton Archive / Getty Images; Chapter 9: © Dennis Kunkel Microscopy, Inc. / Visuals Unlimited / Corbis; Chapter 10: © Bettmann / Corbis; Chapter 11: © Joseph Scherschel / Time & Life Pictures / Getty Images; Chapter 12: © Leemage / Universal Images Group / Getty Images; Chapter 13: © Foodfolio / the food passionates / Corbis; Chapter 14: © Steven Puetzer / The Image Bank / Getty Images; Chapter 15: © Stewart Bremner / Flickr / Getty Images; Chapter 16: © John Wilkes Studio / Corbis; Chapter 17: Netter illustration from www.netterimages.com, used with permission of Elsevier Inc. All rights reserved.

For information about permission to reproduce selections from this book, write to Permissions, W. W. Norton & Company, Inc., 500 Fifth Avenue, New York, NY 10110

For information about special discounts for bulk purchases, please contact W. W. Norton Special Sales at specialsales@wwnorton.com or 800-233-4830

*Manufacturing by Courier Westford*
*Book design by Judith Stagnitto / Abbate Design*
*Production manager: Devon Zahn*

Library of Congress Cataloging-in-Publication Data
Roach, Mary.
Gulp : adventures on the alimentary canal / Mary Roach. — First edition.
   pages cm
Includes bibliographical references.
ISBN 978-0-393-08157-2 (hardcover)
1. Digestive organs—Popular works. 2. Alimentary canal—Popular works. 3. Gastrointestinal system—Popular works. I. Title.
QP145.R53 2013
612.3—dc23

                                                                    2012050391


ISBN 978-0-393-34874-3 pbk.


W. W. Norton & Company, Inc.
500 Fifth Avenue, New York, N.Y. 10110
www.wwnorton.com

W. W. Norton & Company Ltd.
Castle House, 75/76 Wells Street, London W1T 3QT

1 2 3 4 5 6 7 8 9 0

*For Lily and Phoebe,*
*and my brother Rip*

# Contents

# *Gulp*

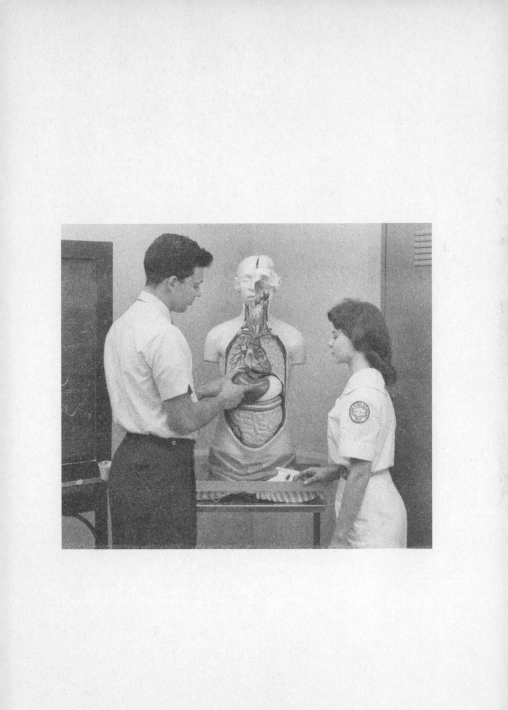

# Introduction

IN 1968, on the Berkeley campus of the University of California, six young men undertook an irregular and unprecedented act. Despite the setting and the social climate of the day, it involved no civil disobedience or mind-altering substances. Given that it took place in the nutritional sciences department, I cannot even say with confidence that the participants wore bell-bottomed pants or sideburns of unusual scope. I know only the basic facts: the six men stepped inside a metabolic chamber and remained for two days, testing meals made from dead bacteria.

This was the fevered dawn of space exploration; NASA had Mars on its mind. A spacecraft packed with all the food necessary for a two-year mission would be impracticably heavy to launch. Thus there was a push to develop menu items that could be "bioregenerated," that is to say, farmed on elements of the astronauts' waste. The title of the paper nicely sums the results: "Human Intolerance to Bacteria as Food." Leaving aside the vomiting and vertigo, the thirteen bowel movements in twelve hours from Subject H, one hopes the aesthetics alone would have tabled further research. Pale gray *Aerobacter*, served as a "slurry," was reported to be unpleasantly slimy. *H. eutropha* had a "halogen-like taste."

Some in the field looked askance at the work. I found this quote in a chapter on fabricated space foods: "Men and women . . . do not ingest nutrients, they consume food. More than that, they . . . eat meals. Although to the single-minded biochemist or physiologist, this aspect of human behavior may appear to be irrelevant or even frivolous, it is nevertheless a deeply ingrained part of the human situation."

The point is well taken. In their zeal for a solution, the Berkeley team would appear to have lost a bit of perspective. When you can identify the taste of street lighting, it may be time to take a break from experimental nutrition. But I wish to say a word in defense of the "single-minded biochemist or physiologist." As a writer, I live for these men and women, the scientists who tackle the questions no one else thinks—or has the courage—to ask: the gastric pioneer William Beaumont, with his tongue through the fistulated hole in his houseboy's stomach; the Swedish physician Algot Key-Åberg, propping cadavers in dining room chairs to study their holding capacity; François Magendie, the first man to identify the chemical constituents of intestinal gas, aided in his investigation by four French prisoners guillotined in the act of digesting their last meal; David Metz, the Philadelphia dyspepsia expert who shot X-ray footage of a competitive eater downing hotdogs two at a time, to see what it might reveal about indigestion; and, of course, our Berkeley nutritionists, spooning bacteria onto dinnerware and stepping back like nervous chefs to see how it goes. The meals were a flop, but the experiment, for better or worse, inspired this book.

When it comes to literature about eating, science has been a little hard to hear amid the clamor of cuisine. Just as we adorn sex with the fancy gold-leaf filigree of love, so we dress the need for sustenance in the finery of cooking and connoisseurship. I adore the writings of M. F. K. Fisher and Calvin Trillin, but I adore no less Michael Levitt ("Studies of a Flatulent Patient"), J. C. Dalton ("Experimental Investigations to Determine Whether the Garden Slug Can Live in the Human Stomach"), and P. B. Johnsen ("A Lexicon of Pond-Raised Catfish Flavor Descriptors"). I'm not saying I don't appreciate a nice meal. I'm saying that the human equipment—and the delightful, unusual people who study it— are at least as interesting as the photogenic arrangements we push through it.

Yes, men and women eat meals. But they also ingest nutrients. They grind and sculpt them into a moistened bolus that is delivered, via a stadium wave of sequential contractions, into a self-kneading sack of hydrochloric acid and then dumped into a tubular leach field, where it is converted into the most powerful taboo in human history. Lunch is an opening act.

*M*Y INTRODUCTION TO human anatomy was missing a good deal of its own. It took the form of a headless, limbless molded-plastic torso* in Mrs. Claflin's science classroom. The chest and rib cage were sheared away, as if by some unspeakable industrial accident, leaving a set of removable organs in full and lurid view. The torso stood on a table in the back of the room, enduring daily evisceration and reassembly at the hands of fifth graders. The idea was to introduce young minds to the geography of their own interior, and at this it failed terribly. The organs fit together like

---

* Similar products exist to this day, under names like "Dual Sex Human Torso with Detachable Head" and "Deluxe 16-Part Human Torso," adding an illicit serial-killer, sex-crime thrill to educational supply catalogues.

puzzle pieces, tidy as wares in a butcher's glass case.* The digestive tract came out in parts, esophagus separate from stomach, stomach from intestines. A better teaching tool would have been the knitted digestive tract that made the rounds of the Internet a few years ago: a single tube from mouth to rectum.

*Tube* isn't quite the right metaphor, as it implies a sameness throughout. The tract is more of a railroad flat: a long structure, one room opening onto the next, though each with a distinctive look and purpose. Just as you would never mistake kitchen for bedroom, you would not, from the perspective of a tiny alimentary traveler, mistake mouth for stomach for colon.

I have toured the tube from that tiny traveler's perspective, by way of a pill cam: an undersized digital camera shaped like an oversized multivitamin. A pill cam documents its travels like a teenager with a smartphone, grabbing snapshots second by second as it moves along. Inside the stomach, the images are murky green with bits of drifting sediment. It's like footage from a *Titanic* documentary. In a matter of hours, acids, enzymes, and the stomach's muscular churning reduce all but the most resilient bits of food (and pill cams) to a gruel called chyme.

Eventually even a pill cam is sent on down the line. As it breaches the pylorus—the portal from the stomach to the small intestine—the décor changes abruptly. The walls of the small

---

* In reality, guts are more stew than meat counter, a fact that went underappreciated for centuries. So great was the Victorian taste for order that displaced organs constituted a medical diagnosis. Doctors had been misled not by plastic models, but by cadavers and surgical patients—whose organs ride higher because the body is horizontal. The debut of X-rays, for which patients sit up and guts slosh downward, spawned a fad for surgery on "dropped organs"—hundreds of body parts needlessly hitched up and sewn in place.

intestine are baloney pink and lush with millimeter-long projections called villi. Villi increase the surface area available for absorbing nutrients. They are the tiny loops on the terry cloth. The inside surface of the colon, by contrast, is shiny-smooth as Cling Wrap. It would not make a good bath towel. The colon and rectum—the farthest reaches of the digestive tract—are primarily a waste-management facility: they store it, dry it out.

Function was not hinted at in Mrs. Claflin's educational torso man. Interior surfaces were hidden. The small intestine and colon were presented as a single fused ravelment, like a brain that had been thrown against the wall. Yet I owe the guy a debt of thanks. To venture beyond the abdominal wall, even a plastic one, was to pull back the curtain on life itself. I found it both appalling and compelling, all the more so because I knew a parallel world existed within my own pinkish hull. I mark that fifth-grade classroom as the point at which curiosity began to push aside disgust or fear or whatever it is that so reliably deflects mind from body.

The early anatomists had that curiosity in spades. They entered the human form like an unexplored continent. Parts were named like elements of geography: the isthmus of the thyroid, the isles of the pancreas, the straits and inlets of the pelvis. The digestive tract was for centuries known as the alimentary canal. How lovely to picture one's dinner making its way down a tranquil, winding waterway, digestion and excretion no more upsetting or off-putting than a cruise along the Rhine. It's this mood, these sentiments—the excitement of exploration and the surprises and delights of travel to foreign locales—that I hope to inspire with this book.

It may take some doing. The prevailing attitude is one of disgust. There are people, anorexics, so repulsed by the thought of their food inside them that they cannot bring themselves to eat.

In Brahmin Hindu tradition, saliva is so potent a ritual pollutant that a drop of one's own spittle on the lips is a kind of defilement. I remember, for my last book, talking to the public-affairs staff who choose what to stream on NASA TV. The cameras are often parked on the comings and goings of Mission Control. If someone spots a staffer eating lunch at his desk, the camera is quickly repositioned. In a restaurant setting, conviviality distracts us from the biological reality of nutrient intake and oral processing. But a man alone with a sandwich appears as what he is: an organism satisfying a need. As with other bodily imperatives, we'd rather not be watched. Feeding, and even more so its unsavory correlates, are as much taboos as mating and death.

The taboos have worked in my favor. The alimentary recesses hide a lode of unusual stories, mostly unmined. Authors have profiled the brain, the heart, the eyes, the skin, the penis and the female geography, even the hair,* but never the gut. The pie hole and the feed chute are mine.

Like a bite of something yummy, you will begin at one end and make your way to the other. Though this is not a practical health book, your more pressing alimentary curiosities will be addressed. And some less pressing. Could thorough chewing lower the national debt? If saliva is full of bacteria, why do animals lick their wounds? Why don't suicide bombers smuggle

---

* *The Hair*, by Charles Henri Leonard, published in 1879. It was from Leonard that I learned of a framed display of presidential hair, currently residing in the National Museum of American History and featuring snippets from the first fourteen presidents, including a coarse, yellow-gray, "somewhat peculiar" lock from John Quincy Adams. Leonard, himself moderately peculiar, calculated that "a single head of hair of average growth and luxuriousness in any audience of two hundred people will hold supported that entire audience" and, I would add, render an evening at the theater so much the more memorable.

bombs in their rectums? Why don't stomachs digest themselves? Why is crunchy food so appealing? Can constipation kill you? Did it kill Elvis?

You will occasionally not believe me, but my aim is not to disgust. I have tried, in my way, to exercise restraint. I am aware of the website www.poopreport.com, but I did not visit. When I stumbled on the paper "Fecal Odor of Sick Hedgehogs Mediates Olfactory Attraction of the Tick" in the references of another paper, I resisted the urge to order a copy. I don't want you to say, "This is gross." I want you to say, "I thought this would be gross, but it's really interesting." Okay, and maybe a little gross.

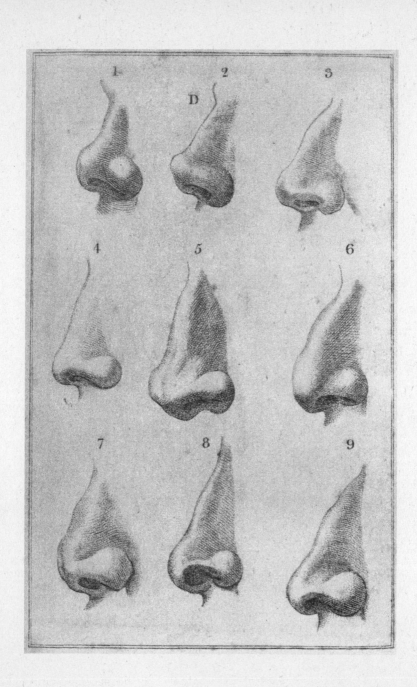

# 1

# *Nose Job*

## TASTING HAS LITTLE TO DO WITH TASTE

**T**HE SENSORY ANALYST rides a Harley. There are surely many things she enjoys about traveling by motorcycle, but the one Sue Langstaff mentions to me is the way the air, the great and odorous out-of-doors, is shoved into her nose. It's a big, lasting passive sniff.* This is why dogs stick their heads out the car window. It's not for the feeling of the wind in their hair. When you have a nose like a dog has, or Sue Langstaff, you take in the

---

* A few words on sniffing. Without it (or a Harley), you miss all but the most potent of smells going on around you. Only 5 to 10 percent of air inhaled while breathing normally reaches the olfactory epithelium, at the roof of the nasal cavity.

Olfaction researchers in need of a controlled, consistently sized sniff use an olfactometer to deliver "odorant pulses." The technique replaces the rather more vigorous "blast olfactometry" as well as the original olfactometer, which connected to a glass and aluminum box called the "camera inodorata." ("The subject's head was placed in the box," wrote the inventor, alarmingly, in 1921.)

sights by smell. Here is California's Highway 29 between Napa and St. Helena, through Langstaff's nose: cut grass, diesel from the Wine Train locomotive, sulfur being sprayed on grapes, garlic from Bottega Ristorante, rotting vegetation from low tide on the Napa River, toasting oak from the Demptos cooperage, hydrogen sulfide from the Calistoga mineral baths, grilling meat and onions from Gott's drive-in, alcohol evaporating off the open fermenters at Whitehall Lane Winery, dirt from a vineyard tiller, smoking meats at Mustards Grill, manure, hay.

Tasting—in the sense of "wine-tasting" and of what Sue Langstaff does when she evaluates a product—is mostly smelling. The exact verb would be *flavoring*, if that could be a verb in the same way *tasting* and *smelling* are. Flavor is a combination of taste (sensory input from the surface of the tongue) and smell, but mostly it's the latter. Humans perceive five tastes—sweet, bitter, salty, sour, and umami (brothy)—and an almost infinite number of smells. Eighty to ninety percent of the sensory experience of eating is olfaction. Langstaff could throw away her tongue and still do a reasonable facsimile of her job.

Her job. It is a kind of sensory forensics. "People come to me and say, 'My wine stinks. What happened?'" Langstaff can read the stink. Off-flavors—or "defects," in the professional's parlance—are clues to what went wrong. An olive oil with a flavor of straw or hay suggests a problem with desiccated olives. A beer with a "hospital" smell is an indication that the brewer may have used chlorinated water, even just to rinse the equipment. The wine flavors "leather" and "horse sweat" are tells for the spoilage yeast *Brettanomyces*.

The nose is a fleshly gas chromatograph. As you chew food or hold wine in the warmth of your mouth, aromatic gases are set free. As you exhale, these "volatiles" waft up through the posterior nares—the internal nostrils* at the back of the mouth—and connect with olfactory receptors in the upper reaches of the nasal cavity. (The technical name for this internal smelling is retronasal olfaction. The more familiar sniffing of aromas through the external nostrils is called orthonasal olfaction.) The information is passed on to the brain, which scans for a match. What sets a professional nose apart from an everyday nose is not so much its sensitivity to the many aromas in a food or drink, but the ability to tease them apart and identify them.

Like this: "Dried cherries. Molasses—blackstrap." Langstaff is sniffing a strong, dark ale called Noel. We are at Beer Revolution, an amply stocked, mildly skunky† bar in Oakland, California, where I have an office (in the city, not the bar) and Langstaff has a parent in the hospital. She could use a drink, and we have four. For demonstration purposes.

In general, Langstaff isn't a talky person. Her sentences present in low, unhurried tones without italics or exclamation points. The question "Which beer do you want, Mary?" went down at the end. When she puts her nose to a glass, though, something switches on. She sits straighter and her words come out faster, lit by interest and focus. "It smells like a campfire to me also. Smokey,

---

* An Internet search on the medical term for nostrils produced this: "Save on Nasal Nares! Free 2-day Shipping with Amazon Prime." They really are taking over the world.

† "Skunky" is between "rotten egg" and "canned corn" on the Defects Wheel for Beer. (Langstaff designed diagnostic wheels for off-flavors in wine, beer, and olive oil.) In the absence of skunks, a mild rendition of skunkiness is achieved by oxidating beer, that is, exposing it to air, as by spilling it or leaving out half-filled glasses.

like wood, charred wood. Like a cedar chest, like a cigar, tobacco, dark things, smoking jackets." She sips from the glass. "Now I'm getting the chocolate in the mouth. Caramel, cocoa nibs . . ."

I sniff the ale. I sip it, push it around my mouth, draw blanks. I can tell it's intense and complex, but I don't recognize any of the components of what I'm experiencing. Why can't I do this? Why is it so hard to find words for flavors and smells? For one thing, smell, unlike our other senses, isn't consciously processed. The input goes straight to the emotion and memory centers. Langstaff's first impression of a scent or flavor may be a flash of color, an image, a sense of warm or cool, rather than a word. Smoking jackets in a glass of Noel, Christmas trees in a hoppy, resinous India pale ale.

It's this too: Humans are better equipped for sight than for smell. We process visual input ten times faster than olfactory. Visual and cognitive cues handily trump olfactory ones, a fact famously demonstrated in a 2001 collaboration between a sensory scientist and a team of oenologists (wine scientists) at the University of Bordeaux in Talence, France. Fifty-four oenology students were asked to use standard wine-flavor descriptors to describe a red wine and a white wine. In a second round of tasting, the same white wine was paired with a "red," which was actually the same white wine yet again but secretly colored red. (Tests were run to make sure the red coloring didn't affect the flavor.) In describing the red-colored white wine, the students dropped the white wine terms they'd used in the first round in favor of red wine descriptors. "Because of the visual information," the authors wrote, "the tasters discounted the olfactory information." They believed they were tasting red wine.

Verbal facility with smells and flavors doesn't come naturally.

As babies, we learn to talk by naming what we see. "Baby points to a lamp, mother says, 'Yes, a lamp,'" says Johan Lundström, a biological psychologist with the Monell Chemical Senses Center in Philadelphia. "Baby smells an odor, mother says nothing." All our lives, we communicate through visuals. No one, with a possible exception made for Sue Langstaff, would say, "Go left at the smell of simmering hotdogs."

"In our society, it's important to know colors," Langstaff says over a rising happy-hour din. We need to know the difference between a green light and a red light. It's not so important to know the difference between bitter and sour, skunky and yeasty, tarry and burnt. "Who cares. They're both terrible. Ew. But if you're a brewer, it's extremely important." Brewers and vintners learn by exposure, gradually honing their focus and deepening their awareness. By sniffing and contrasting batches and ingredients, they learn to speak a language of flavor. "It's like listening to an orchestra," Langstaff says. At first you hear the entire sound, but with time and concentration you learn to break it down, to hear the bassoon, the oboe, the strings.*

---

* In 2010, inventor George Eapen and snack-food giant Frito-Lay took the comparison beyond the realm of metaphor. They patented a system whereby snack-food bags could be printed with a bar code allowing consumers to retrieve and download a fifteen-second audio clip of a symphonic interlude, with the different instruments representing the various flavor components. Eapen, in his patent, used the example of a salsa-flavored corn chip. "A piano intro begins upon the customer's perception of the cilantro flavoring. . . . The full band section occurs at approximately the time that the consumer perceives the tomatillo and lime flavors. . . . A second melody section corresponds to the sensation of the heat burn imparted by the Serrano chili." U.S. Patent No. 7,942,311 includes sheet music for the salsa-flavored chip experience.

As with music, some people seem born to it. Maybe they have more olfactory receptors or their brain is wired differently, maybe both. Langstaff liked to sniff her parents' leather goods as a small child. "Purses, briefcases, shoes," she says. "I was a weird kid." My wallet is on the table, and without thinking, I stick it under her nose. "Yeah, nice," she says, though I don't see her sniff. The performing-chimp aspect of the work gets tiresome.

While not discounting genetic differences, Langstaff believes sensory analysis is mainly a matter of practice. Amateurs and novices can learn via kits, such as Le Nez du Vin, made up of many tiny bottles of reference molecules: isolated samples of the chemicals that make up the natural flavors.

A quick word about chemicals and flavors. All flavors in nature are chemicals. That's what food is. Organic, vine-ripened, processed and unprocessed, vegetable and animal, all of it chemicals. The characteristic aroma of fresh pineapple? Ethyl 3-(methylthio) propanoate, with a supporting cast of lactones, hydrocarbons, and aldehydes. The delicate essence of just-sliced cucumber? $2E,6Z$-Nonadienal. The telltale perfume of the ripe Bartlett pear? Alkyl $(2E,4Z)$-2,4-decadienoates.

$O$F THE FOUR half-pints on the table between us, Langstaff prefers the lightest, a strawberry wheat beer. I like the IPA best, but to her that's not a "sitting and sipping" beer. It's something she'd drink with food.

I ask Sue Langstaff—sensory consultant to the brewing industry for twenty-plus years, twice a judge at the Great American Beer Festival—what she'd order right now if she had to choose between an IPA and a Budweiser.

"I'd get Bud."

"Sue, no."

"Yes!" First exclamation point of the afternoon. "People pooh-pooh Bud. It's an extremely well-made beer. It's clean, it's refreshing. If you're mowing the lawn and you come in and you want something refreshing and thirst-quenching, you wouldn't drink this." She indicates the IPA.

Of all the descriptors in the Beer Flavor Lexicon I brought with me today, Langstaff would apply just two to Bud: malty and worty. She warns me about equating complexity with quality. "All that stuff you read on wine bottles, in wine magazines, where they throw out a dozen descriptors? That's not sensory evaluation. That's marketing."

Taste—as in personal preference, discernment—is subjective. It's ephemeral, shaped by trends and fads. It's one part mouth and nose, two parts ego. Even flavors that professional evaluators agree are "defects" can come to signify superior taste. Langstaff mentions a small brewery in northern California that has been taking its beers right up to the doorstep of defective, adding strains of bacteria known for their spoilage effects. Whether through exposure or a desire to ride the cutting edge, people can acquire a taste for pretty much anything. If they can come to like the smelly-foot stink of Limburger cheese or the corpsey reek of durian fruit, they can come to enjoy bacteria-soured beer. (One assumes there are limits, however. Leaving olive oil in contact with rotting sediment at the bottom of a tank can create flavors enumerated on Langstaff's Defects Wheel for Olive Oil as follows: "baby diapers, manure, vomit, bad salami, sewer dregs, pig farm waste pond.")

Because it's hard for people to gauge quality by flavor, they tend to gauge it by price. That's a mistake. Langstaff has evaluated wine professionally for twenty years. In her opinion, the difference between a $500 bottle of wine and one that costs $30 is largely hype. "Wineries that sell their wines for $500 a bottle have the same problems as wineries that sell their wine for $10 a bottle. You can't make the statement that if it's low-cost it's not well made." Most of the time, people don't even prefer the expensive bottle— provided they can't see the label. Paul Wagner, a top wine judge and founding contributor to the industry blog Through the Bung-hole, plays a game with his wine-marketing classes at Napa Valley College. The students, most of whom have several years' experience in the industry, are asked to rank six wines, their labels hidden by—a nice touch here—brown paper bags. All are wines Wagner himself enjoys. At least one is under $10 and two are over $50. "Over the past eighteen years, every time," he told me, "the least expensive wine averages the highest ranking, and the most expensive two finish at the bottom." In 2011, a Gallo cabernet scored the highest average rating, and a Chateau Gruaud Larose (which retails from between $60 and $70) took the bottom slot.

Unscrupulous vendors turn the situation to their advantage. In China, nouveau-riche status-seekers are spending small fortunes on counterfeit Bordeaux. A related scenario exists here vis-à-vis olive oil. "The United States is a dumping ground for bad olive oil," Langstaff told me. It's no secret among European manufacturers that Americans have no palate for olive oils. The Olive Center—a recent addition to the Robert Mondavi Institute for Wine and Food Science, on the campus of the University of California at Davis—aims to change that.

It starts with tastings. I don't know which vineyard first ushered wine-tasting off the palates of vintners and into the mouths of everyday consumers, but it was a stroke of marketing genius.

Wine-tastings spawn wine enthusiasts, wine collecting, wine tourism, wine magazines, wine competitions, (wine addictions,)—all of it adding up to a multibillion-dollar industry. Olive trees grow in the same climate and soil conditions as grapes. The olive oil people have been up in Napa Valley all along, going, "Hey, how do we get a piece of this action?"

In addition to hosting tastings, the Olive Center has hired Langstaff to train a new UC Davis Olive Oil Taste Panel. Taste panels (or flavor panels, as they are more accurately called) have typically been made up of industry professionals. Langstaff wants to open it up to novices, for the simple reason that know-nothings are easier to train than know-it-alls. The center has a call for apprentice tasters on its website. The "tryouts" are coming up. At least one know-nothing will be there for sure.


THE OLIVE CENTER is smaller than its name suggests. It consists of a single office and a shared receptionist on the first floor of the Sensory Building at the Robert Mondavi Institute. Bottles of oil and canned olives line the tops of cabinets and have begun to colonize the wall-to-wall. There's no room in the center to hold the tryouts, so they are taking place next door in the Silverado Vineyards Sensory Theater, the building's lecture hall and classroom tasting facility. (Silverado helped fund it. Additionally, each seat has a sponsor, with the name engraved on a small plaque.)

Langstaff makes her entrance burdened like a pack mule. Three tote bags hang off her shoulders, and she wheels a multi-tiered cart crammed with oils, laptops, water bottles, and stacks of cups. She wears dun-colored pants, black sport sandals, and a short-sleeved shirt in the Hawaiian style, though without an island motif. She calls roll: twenty names. Of them, twelve will make the first cut, and six will go on to apprentice.

Langstaff lays out the ground rules for future apprentices: be here, be on time. Be agreeable. "We will be evaluating some nasty oils. You will have to put them in your mouth.* For the good of science. For the good of olive oil. We are here to help the producers, to tell them, What attributes does the oil have, does it have defects, what can they do differently next year—treat the olives better, pick them at a different time, et cetera." There will be no pay. No one will reimburse for the seven-dollar parking-garage fee. The existing panelists are known to have some prickle, to borrow an official olive-oil sensory descriptor.

"You may be thinking, wow, I really don't want to be on this thing." The faint of heart are invited to pack up and go. No one moves.

"All right then." Langstaff surveys the room. "Shields up." She is referring to removable panels used to partition the room's long tables into private tasting booths. This way, you aren't influenced by the facial expressions (or test answers) of the people seated next to you. Hired sensory-science students move along the rows, pull-

---

* It could be worse. In 1984, goat-milk flavor panelists were enlisted by a team of Pennsylvania ag researchers to sleuth the source of a nasty "goaty" flavor that intermittently fouls goat milk. The main suspect was a noxious odor from the scent glands of amorous male goats. But there was also this: "The buck in rut sprays urine over its chin and neck area." Five pungent compounds isolated from the urine and scent glands of rutting males were added, one at a time, to samples of pure, sweet goat milk. The panelists rated each sample for "goaty" "rancid," and "musky-melon" flavors. Simple answers proved elusive. "A thorough investigation of 'goaty' flavor," the researchers concluded, "is beyond the scope of this paper."

ing the panels out of slots in the front of the tables and sliding them into place, like helpers on a game-show set.

A plastic tray is set in front of each of us. The trays hold eight small lidded cups: our first test. Each cup holds an aromatic liquid. Swirl, sniff, identify. A few seem easy: almond extract, vinegar, olive oil. Apricot required two full minutes of deep thought. Others remain unfamiliar no matter how many times and how deeply I sniff. According to the journal *Chemical Senses*, a "typical human sniff" has a duration of 1.6 seconds and a volume of about two cups. I'm sniffing twice as hard. I'm sniffing the way clueless Americans try to make non-English speakers understand them by shouting. One aroma will turn out to be olive brine—the water from a bottle or can of olives. Reflecting the preponderance of olive people trying out today, an impressive thirteen out of twenty get this right.

Next is a "triangle test": three olive-oil samples, two of them identical. Our task is to identify the odd one out. We are given paper cups of water for rinsing and, for spitting, large red plastic cups of the kind that litter the lawns and porches of frat houses on weekend mornings. The red here today perhaps serving as a warning: Do not drink! Langstaff sits at the front of the room, reading a newspaper.

It's not going well here in the B.R. Cohn Winery seat. All three oils taste the same to me: a hint of freshly mown grass, with a peppery finish. I do not detect apple, avocado, melon, pawpaw, old fruit bowl, almond, green tomato, artichoke, cinnamon, cat urine, hemp, Parmesan cheese, fetid milk, Band-Aid, crushed ants, or any other olive-oil flavor, good or bad, that might set one of these oils apart. With time running out, I don't bother spitting. I'm sipping oil like it's tea. Langstaff glances at me over her glasses. I wipe my lips and chin with my palm, and a shiny smear comes away.

Our final challenge is a ranking test: five olive oils of differing degrees of bitterness. This proves a challenge for me, as I would not have described any of them as bitter. All around me, people make sounds like ill-mannered soup-eaters, aerating the oils to free the aromatic gases. I'm doing a *mnyeh-mnyeh-mnyeh* Bugs Bunny thing with my tongue, but it's not helping. Well before the test period ends, I stop. I do something I've never done in my entire overachieving life. I give up and guess. I do this partly at the behest of my stomach, which is struggling to cope with the unusual delivery of a sizable amount of straight olive oil.

After everyone else leaves, Langstaff shares some of the group's answers (with names removed). Those who performed well on the oil rankings—incredibly, several got it close to exact—also noted that aroma number 7, on the first test, was not just olive oil, but *rancid* olive oil. Four out of twenty people, all olive professionals, nailed that detail. (The oil smelled fine to me. I was right there with the numb-nose who wrote, on his answer form, "Oh, for a piece of good bread!")

Here's what I find interesting. The people who work with olives and olive oil, most of whom performed supernaturally well on the ranking and triangle tests, were occasionally stumped by some of the most common and, to me, obvious aromas. A woman who, in the initial sniff test, realized that the olive oil was "rancid, fusty" failed to recognize almond extract. She wrote, "Cranberry, fruity, sweet, aloe juice." She described diacetyl, the smell of artificial (movie popcorn) butter, as "licorice, candy, bubble gum." Those aren't important flavors in the day-to-day of the olive world, so there's no reason for her to know them. This supports what Langstaff said earlier. As with any language, proficiency builds with exposure and practice. (Though not quickly; the average training period for a sensory panelist is sixty hours.)

In my case, it won't be happening any time soon. An e-mail from Langstaff arrives around nine that night. "Hi Mary. Hope you enjoyed the tryouts. Unfortunately you did not make the cut."

SENSORY ANALYSIS IS not limited to the epicurean industries of Napa Valley. For any food or drink manufactured on a reasonably large scale, there are trained panelists and sensory descriptors. Poking around in the sensory-science journals, I have seen flavor lexicons for mutton, strawberry yogurt, chicken nuggets, ripening anchovies, almonds, beef, chocolate ice cream, pond-raised catfish, aged Cheddar cheese, rice, apples, rye bread, and "warmed-over flavor."

The work entails more than just troubleshooting. Sensory analysts and panels help with product development. They keep the flavors of established products on track when a formula is altered— say, to lower the fat or salt content. They work with the market research staff. When focus groups of consumers prefer one version of, say, a ranch dressing over another (or over a competitor's dressing), sensory evaluators may be brought in to figure out the salient attributes of the more popular item. The food scientists can then work backward from those attributes to tweak the formula.

Why use humans rather than lab equipment? Because the latter would yield dozens of chemical differences* between a pair of products. Without a human evaluator, it's impossible to assign

---

* Probably more. The *Handbook of Fruit and Vegetable Flavors* includes a four-page table of aroma compounds identified in fresh pineapple: 716 chemicals in all.

sensory meaning to them. Which of those dozens of differences in chemical makeup translates to a perceptible flavor shift, and which is below the threshold for human detection? Which ones, in short, make the difference in the consumer's mouth and mind? "And you can't ask the consumer," says Langstaff. "You ask the consumer, 'Why does it taste better?' They say, 'Because I like it better.'" The consumer's flavor lexicon is tiny: yum and yuck.

Which product the sensory evaluator prefers, by the way, is irrelevant. He or she may not like any of them, or even the general category. (Langstaff, for instance, rarely drinks beer for pleasure.) "You don't ask your gas chromatograph if it likes the olive oil it's analyzing," Langstaff told us at the tryouts. The goal is to be as neutral, as analytical—as "Mr. Spock"—as possible.

This perhaps explains how it was possible for a team of Canadian researchers to find nine men and women willing to create a canned-cat-food flavor lexicon and a set of tasting protocols. For humans. Tasting cat food. And they couldn't be shy about it. The protocol for evaluating the "meat chunk" portion ("gravy gel" having its own distinct protocol) stipulated that the sample be "moved around mouth and chewed for 10 to 15 seconds, [and] a portion of the sample swallowed."

The idea was to come up with a sort of code, a way to translate the mute preferences of cats. In theory, companies could use human tasters and sensory profiles of the foods cats like in order to predict the success of new formulations. In practice, the technique never really took off.

Because there was a concern that people with a "strong negative attitude" toward tasting cat food would drop out before the project ended, panel applicants at the initial screening were asked not only to describe the cat foods but also to rate them according to how much they liked them. (The average rating, I am gobsmacked to report, fell between "like mildly" and "neither like nor dislike.") Thanks to this unusual data set, we now know that humans prefer cat food with a tuna or herbal flavor over cat food with the flavor descriptors "rancid," "offaly," "cereal," or "burnt."

But humans, as we are about to see, are not cats.

# 2

# *I'll Have the Putrescine*

YOUR PET IS NOT LIKE YOU

**D**ESPITE THE CRYPTIC name and anonymous office-park architecture, the nature of the enterprise that goes on at AFB International is clear the moment you sit down for a meeting. The conference room smells like kibble. One wall of it, entirely glass, looks onto a small-scale kibble extrusion plant where men and women in lab coats and blue sanitary shoe covers tootle here and there pushing metal carts. AFB makes flavor coatings for dry pet foods. To test the coatings, they first need to make small batches of plain kibble and add the coatings. The flavored kibbles are then presented to consumer panels for feedback. The panelists—Spanky, Thomas, Skipper, Porkchop, Rover, Elvis, Sandi, Bela, Yankee, Fergie, Murphy, Limburger, and some three hundred other dogs and cats—reside at AFB's Palatability Assessment Resource Center (PARC), about an hour's drive from the company's suburban St. Louis headquarters.

AFB Vice President Pat Moeller, myself, and a few other staff members are seated around an oval conference table. Moeller is middle-aged, likable, and plain-spoken. He has a small mouth with naturally deep red lips and a pronounced Cupid's bow, but it would be inaccurate to say he has a feminine appearance. Moeller once consulted for NASA, and he has that look. The fundamental challenge of the pet-food professional, Moeller is saying, is to balance the wants and needs of pets with those of their owners. The two are often at odds.

Dry, cereal-based pet foods caught on during World War II, when tin-rationing put a stop to canning, including the canning of dog food made from horse meat (of which there was an abundance around the time Americans embraced the automobile and began selling their mounts to the knackers). Regardless of what pets made of the change, owners were delighted. Dry pet food was less messy and stinky, and more convenient. As a satisfied Spratt's Patent Cat Food customer of yesteryear put it, the little biscuits were "both handy and cleanly."

To meet pets' nutrition requirements while also giving humans the cheap, handy, cleanly product they demand, mainstream pet-food manufacturers blend animal fats and meals with soy and wheat grains and add vitamins and minerals. This yields a cheap, nutritious pellet that no one wants to eat. Cats and dogs are not grain-eaters by choice, Moeller is saying. "So our task is to find ways to entice them to eat enough for it to be nutritionally sufficient."

This is where "palatants" enter the scene. AFB designs powdered flavor coatings for the edible extruded shapes. Moeller came to AFB from Frito-Lay, where his job was to design, well, powdered flavor coatings for edible extruded shapes. "There are," he allows, "a lot of parallels." A Cheeto without its powdered coating has almost no flavor.* Likewise, the sauces on processed convenience meals are basically palatants for humans. The cooking process for the chicken in a microwavable entrée imparts a mild to nonexistent flavor. The flavor comes almost entirely from the sauce—by design. Says Moeller, "You want a common base that you can put two or three or more different sauces on and have a full product line."

Pet foods come in a variety of flavors because that's what we humans like,† and we assume our pets like what we like. We have that wrong. "For cats especially," Moeller says, "change is often more difficult than monotony."

Nancy Rawson, seated across from me, is AFB's director of basic research and an expert in animal taste and smell. She volunteers that cats are more or less "monoguesic," meaning they stick to one food. Outdoor cats tend to be either mousers or birders, not both. But don't worry, as most of the difference between Tuna Treat and Poultry Platter is in the name and the picture on the label. "They may have more fish meal in one and more poultry meal in another," says Moeller, "but the flavors may or may not change."

The extent to which Americans project their own food qualms and biases onto their pets has lately veered off into the absurd. Some of AFB's clients have begun marketing 100 percent vegetar-

---

* Moeller, who has tasted the naked Cheeto, likens it to a piece of unsweetened puffed corn cereal.

† Or that's what we think we like. In reality, the average person eats no more than about thirty foods on a regular basis. "It's very restricted," says Adam Drewnowski, director of the University of Washington Center for Obesity Research, who did the tallying. Most people ran through their entire repertoire in four days.

ian kibble for cats. The cat is what's called a true carnivore; its natural diet contains no plants.

Moeller tilts his head. A slight lift of the eyebrows. The look says, "Whatever the client wants."

NANCY RAWSON KNOWS how to get a cat to finish its vegetables. Pyrophosphates have been described to me as "cat crack." Coat some kibble with it, and you, the pet-food manufacturer, can make up for a whole host of gustatory shortcomings. Rawson has three kinds of pyrophosphate in her office. They're in plain brown-glass bottles, vaguely sinister in their anonymity. I asked to try them, which, I think, has won me some points. Sodium acid pyrophosphate, known affectionately as SAPP, is part of the founding patent for AFB, yet almost no one who works for the company has ever asked to taste it. Rawson finds this odd. I do too, though I also accept the possibility that other people would find the two of us odd.

Rawson is dressed today in a floral-print skirt, on the long side, with low-heeled brown boots and a lightweight plum-colored sweater. She is tall and thin, with wide, graceful cheek and jaw bones. She looks at once like someone who could have worked as a runway model and someone who would be mildly put off to hear that. She is brainy and hard working, committed to her job in a way you don't necessarily expect pet-food people to be. Before she was hired at AFB, she was a nutritionist at Campbell's Soup Company, and before that, she did research on animal taste and smell at the Monell Chemical Senses Center.

Rawson unscrews the cap of one of the bottles. She pours a finger of clear liquid into a plastic cup. Though pet-food palatants most often take the form of a powder, liquid is better for tasting.

To experience taste, the molecules of the tastant—the thing one is tasting—need to dissolve in liquid. Liquid flows into the microscopic canyons of the tongue's papillae, coming into contact with the "buds" of taste receptor cells that cover them. That's one reason to be grateful for saliva. Additionally, it explains the appeal of dunking one's doughnuts.

Taste is a sort of chemical touch. Taste cells are specialized skin cells. If you have hands for picking up foods and putting them into your mouth, it makes sense for taste cells to be on your tongue. But if, like flies, you don't, it may be more expedient to have them on your feet. "They land on something and go, 'Oooo, sugar!'" Rawson does her best impersonation of a housefly. "And the proboscis automatically comes out to suck the fluids." Rawson has a colleague who studies crayfish and lobsters, which taste with their antennae. "I was always jealous of people who study lobsters. They examine the antennae, and then they have a lobster dinner."

The study animal of choice for taste researchers is the catfish,* simply because it has so many receptors. They are all over its skin. "Catfish are basically swimming tongues," says Rawson. It is a useful adaptation for a limbless creature that locates food by brushing up against it; many catfish species feed by scavenging debris on the bottom of rivers.

I try to imagine what life would be like if humans tasted things by rubbing them on their skin. *Hey, try this salted caramel gelato, it's amazing.* Rawson points out that a catfish may not consciously perceive anything when it tastes its food. The catfish neurological system may simply direct the muscles to eat. It seems

---

* This explains the perplexing odor of swamp water on certain floors of the Monell Chemical Senses Center during the 1980s. The basement was a big catfish pond.

odd to think of tasting without any perceptive experience, but you may be doing it right now. Humans have taste receptor cells in the gut, the voice box, the upper esophagus, but only the tongue's receptors report to the brain. "Which is something to be thankful for," says Danielle Reed, Rawson's former colleague at Monell. Otherwise you'd be tasting things like bile and pancreatic enzymes. (Intestinal taste receptors are thought to trigger hormonal responses to molecules, such as salt and sugar, and defensive reactions—vomiting, diarrhea—to dangerous bitter items.)

We consider tasting to be a hedonic pursuit, but in much of the animal kingdom, as well as in our own prehistory, the role of taste was more functional than sensual. Taste, like smell, is a doorman for the digestive tract, a chemical scan for possibly dangerous (bitter, sour) elements and desirable (salty, sweet) nutrients. Not long ago, a whale biologist named Phillip Clapham sent me a photograph that illustrates the consequences of life without a doorman. Like most creatures that swallow their food whole, sperm whales have a limited-to-nonexistent sense of taste. The photo is a black-and-white still life of twenty-five objects recovered from sperm whale stomachs. It's like Jonah set up housekeeping: a pitcher, a cup, a tube of toothpaste, a strainer, a wastebasket, a shoe, a decorative figurine.

Enough stalling. Time to try the palatant. I raise the cup to my nose. It has no smell. I roll some over my tongue. All five kinds of taste receptors stand idle. It tastes like water spiked with strange. Not bad, just other. Not food.

"It may be that that otherness is something specific to the cat," says Rawson. Perhaps some element of the taste of meat that humans cannot perceive. The feline passion for pyrophosphates might explain the animal's reputation as a picky eater. "We make [pet food] choices based on what we like," says Reed, "and then when they don't like it, we call them finicky."

There is no way to know or imagine what the taste of pyro-phosphate is like for cats. It's like a cat trying to imagine the taste of sugar. Cats, unlike dogs and other omnivores, can't taste sweetness. There's no need, since the cat's diet in the wild contains almost nothing in the way of carbohydrates (which include simple sugars). Either cats never had the gene for detecting sweet, or they lost it somewhere down the evolutionary road.

Rodents, on the other hand, are slaves to sweetness. They have been known to die of malnutrition rather than step away from a sugar-water drip. In an obesity study from the 1970s, rats fed an all-you-can-eat "supermarket" diet that included marshmallows, milk chocolate, and chocolate-chip cookies gained 269 percent more weight than rats fed standard laboratory fare. There are strains of mice that will, over the course of a day, consume their own bodyweight in diet soda, and you do not want the job of changing their bedding.

Does that mean rodents feel pleasure in tasting sweet things the same way we do? Or is it simply a sequence of programmed responses, receptors sending signals and signals driving muscles? Video footage Danielle Reed sent me suggests that rodents do consciously perceive and savor the taste of something sweet. One clip shows a white mouse that has just been drinking sugar solution. She is shown in ultra-slow motion, filmed from below through a clear plastic floor, licking the fur around the sides of her mouth. (The caption uses the scientific term for lip-licking: "lateral tongue protrusion.") Another clip shows a mouse that has just tasted denatonium benzoate, a bitter compound that parents used to paint on their children's fingertips to discourage nail-biting. The mouse is doing everything it can to rid itself of traces of the chemical. It shakes its head and rubs its face with its hairy white forelegs. It pulls a "gape": mouth opened wide, tongue stuck out to eject the offending food. (Humans do this too. The scientific term here is "the disgust face.")

"If it's exceedingly nasty," Reed told me, "they will actually drag their tongue on the bedding to try to get it off." Clearly taste matters to them.

Conversely, do animals with no taste buds derive no pleasure from eating? Is it just a daily chore? Has anyone observed—in, say, a python eating a rat—those same parts of the brain that light up when humans are experiencing taste delight? Reed doesn't know. "But no doubt somewhere in the world there's a scientist trying to get a live python into an fMRI machine."

Rawson points out that although snakes can't taste, they have a primitive sense of smell. They'll extend their tongue to gather volatile molecules and then pull it back in and plug it into the vomeronasal organ at the roof of the mouth to get a reading. Snakes are keenly attuned to the aroma of favored prey—so much so that if you slip a rat's face and hide, Hannibal Lecter–style, over the snout of a non-favored prey item, a python will try to swallow it. (University of Alabama snake digestion expert Stephen Secor did this some years back to reenact a scene for National Geographic television. "Worked like a charm," he told me. "I can get a python to eat a beer bottle if I put a rat head on it.")

For part of their development, human fetuses have a vomeronasal organ, though no one knows whether it's functional. You can no more ask a fetus about these things than a python. Rawson surmises that the organ is a holdover from "when we were crawling out of the primordial soup,* and we needed to sense the chemicals in the environment and know which ones to go toward or away from."

---

* Not a Campbell's product.

Rawson has an idea of what it is like to eat without perceiving tastes, because she has talked to cancer patients whose taste receptors have been destroyed by radiation treatments. The situation is well beyond unpleasant. "Your body is saying, 'It's not food, it's cardboard,' and it won't let you swallow. No matter how much you tell your brain that you need to eat to survive, you'll gag. These people can actually die of starvation." Rawson knows a researcher who has been experimenting with using potent flavors—which, as we know from the last chapter, are mainly smells—to make up for absent tastes. Taste and smell are intertwined in ways we don't consciously appreciate. Food technologists sometimes exploit the synergy between the two. By adding strawberry or vanilla—aromas we associate with sweetness—it's possible to fool people into thinking a food is sweeter than it really is. Though sneaky, this is not necessarily bad, because it means the product can contain less added sugar.

Which takes us back to palatants, and why pet-food manufacturers love them. As one AFB employee put it, "The client can go, 'Here's my product. I want to cut corners here and here and here, and I want you to cover up all the sins.'" This is especially doable with dog food, as dogs rely more on smell than taste in making choices about what to eat and how vigorously. (Pat Moeller estimates that for dogs, the ratio for how much aroma matters to how much taste matters is 70/30. For cats, the ratio is more like 50/50.) The takeaway lesson is that if the palatant smells appealing, the dog will dive in with instant and obvious zeal, and the owner will assume the food is a hit. In reality it may have only smelled like a hit.

Interpreting animals' eating behaviors is tricky. By way of example, one of the highest compliments a dog can pay its food is to vomit. When a "gulper," to use Pat Moeller's terminology, is excited by the aroma of a food, it will wolf down too much too fast. The stomach overfills, and the meal is reflexively sent back up to avoid any chance of a rupture. "No consumer likes that, but it's the best indication that the dog just loved it." Fortunately for the staff at the AFB Palatability Assessment Resource Center, there are other ways to gauge a pet food's popularity.

"*E*VERYONE WANTS TO be Meow Mix." Amy McCarthy, head of PARC, stands outside the plate-glass window of Tabby Room 2, where an unnamed client is facing off against Meow Mix, Friskies, and uncoated kibble in a preference test. If a client wants to be able to say that cats prefer its product over Meow Mix, they must prove it at a facility like PARC.

Two animal techs dressed in tan surgical scrubs stand facing each other. They hold shallow metal pans of kibble in various shades of brown,* one in each hand. Around their ankles, twenty cats mince and turn. The techs sink in tandem to one knee, lowering the pans.

The difference between dog and cat is immediately obvious. While a dog almost (and occasionally literally) inhales its food the moment it's set down, cats are more cautious. A cat wants to taste a little first. McCarthy directs my gaze to the kibble that has no palatant coating. "See how they feel it in their mouth and then drop it?"

I see an undifferentiated ground-cover of bobbing cat heads, but nod anyway.

---

* Gone are the colored pet-food pieces of the early 1990s. "Because when it comes back up, then you have green and red dye all over your carpet," says Rawson. "That was a huge *duh*."

"Now look there." She directs my gaze to the Meow Mix, where the bottom of the pan is visible through an opening in the kibble. I ask McCarthy if there's an industry term* for the open spot.

"Um . . . 'The space where kibble used to be'?" McCarthy speaks louder than you expect a person to, perhaps a side effect of time spent talking over barking. She is in her thirties, with blonde hair that is center-parted and wants to fall in her face. Every few minutes, she'll raise both forefingers to the sides of her face to nudge it back. Rawson's hair, by contrast, is cropped close to her head. It's a "pixie cut," but those probably aren't the words she used when she discussed it with her haircutter. Rawson has come with me to PARC because she hasn't yet visited and wants to learn how the preference testing is being done and how the techniques might be improved.

Meanwhile, down the hallway, dog kibble A, dressed in a coat of newly formulated AFB palatant, is up against the competitor. The excitement is audible. One dog squeals like sneaker soles on a basketball court. Another makes a huffing sound reminiscent of a two-man timber saw. The techs are wearing heavy-duty ear protection, the kind worn on airport tarmacs.

A tech named Theresa Kleinsorge opens the door of a large kennel crate and sets down two bowls in front of a terrier mix with dark-ringed eyes. Theresa is short and brassy, with spiky magenta-dyed hair. *Kleinsorge* is German for "little trouble," and it seems like a good name—*trouble* in the affectionate sense of well-intentioned mischief. She owns seven dogs. Amy McCarthy shares her home with six. Dog love is palpable here at PARC. It is the first

---

* My brother works in market research. One time after he visited I found a thick report in the trash detailing consumers' feelings about pre-moistened towelettes. It contained the term "wiping events."

pet-food test facility to "group-house" its animals. Other than during certain preference tests, when animals are crated to avoid distractions, PARC is a cageless facility. Groups of dogs, matched by energy level, spend their days roughhousing in outdoor yards.

The terrier mix is named Alabama. His tail thumps a beat on the side of the crate. "Alabama is a gobbler real bad," Theresa says. In making their reports, the AFB techs must take into account the animals' individual mealtime quirks. There are gulpers, circlers, tippers, snooters. If you weren't acquainted with Alabama's neighbor Elvis, for example, you'd think he was blasé about both foods just now set before him. Theresa gives a running commentary of Elvis's behavior while a colleague jots notes. "Sniffing A. Sniffing B. Licking B, licking his paws. Going back to A. Looking at A. Sniffing B. Eating B."

Most dogs are more decisive. Like Porkchop. "You'll see. He'll sniff both, pick one, eat it. Ready?" She puts two bowls by Porkchop's front paws. "Sniffing A, sniffing B, eating A. See? That's what he does."

PARC techs also try to keep a bead on doggy interactions in the yards. "We need to know," says McCarthy. "'Are you down because you don't like the food or because Pipes stole your bone earlier?'" Theresa volunteers that a dog named Rover has lately had a stomach upset, and Porkchop likes to eat the vomit. "So that's cutting into Porkchop's appetite." And probably yours.

In addition to calculating how much of each food the dogs ate, PARC techs tally the first-choice percentage: the percentage of dogs who stuck their snout in the new food first. This is important to a pet-food company because with dogs, as Moeller said earlier, "if you can draw them to the bowl, they'll eat, most of the time." Once the eating begins, though, the dog may move to the other food and wind up consuming more of it. Since most people don't present their dog with two choices, they don't know the extent to which their pet's initial, slavering, scent-driven enthusiasm may have dimmed as the meal went on.

The challenge is to find an aroma that drives dogs wild without making their owners, to use an Amy McCarthy verb, yack. "Cadaverine is a really exciting thing for dogs," says Rawson. "Or putrescine." But not for humans. These are odoriferous compounds given off by decomposing protein. I was surprised to learn that dogs lose interest when meat decays past a certain point. It is a myth that dogs will eat anything. "People think, Dogs love things that are old, nasty, drug around in the dirt," Moeller told me earlier. But only to a point, he says. And for a reason. "Something that's just starting to decay still has full nutritional value. Whereas something where the bacteria have really broken it down, it's lost a lot of its nutritional value and they would only eat it if they had no choice." Either way, a pet owner doesn't want to smell it.

Some dog-food designers go too far in the other direction, tailoring the smell to be pleasing to humans* without taking the dog's experience of it into account. The problem is that the average dog's nose is about a thousand times more sensitive

---

* The Holy Grail is a pet food that not only smells unobjectionable, but also makes the pets' feces smell unobjectionable. It's a challenge because most things you could add to do that will get broken down in digestion and rendered ineffectual. Activated charcoal is problematic because it binds up not just smelly compounds, but nutrients too. Hill's Pet Nutrition experimented with adding ginger. It worked well enough for a patent to have been granted, which must have been some consolation to the nine human panelists tasked with "detecting differences in intensity of the stool odor by sniffing the odor through a port."

than the average human's. A flavor that to you or me is reminiscent of grilling steak may be overpowering and unappealing to a dog.

Earlier in the day, I watched a test of a mint-flavored treat marketed as a tooth-cleaning aid. Chemically speaking, mint, like jalapeño, is less a flavor than an irritant. It's an uncommon choice for a dog treat.* The manufacturers are clearly courting the owners, counting on the association of mint with good oral hygiene. The competition courts the same dental hygiene association but visually: the biscuit is shaped like a toothbrush. Only Rover preferred the minty treats. Which maybe explains the vomiting.

A dog named Winston is nosing through his bowl for the occasional white chunk among the brown. Many of the dogs picked these out first. They're like the M&M's in trail mix. McCarthy is impressed. "That's a really, really palatable piece in there." One of the techs mentions that she tried some earlier, and that the white morsels are chicken. Or rather, "chickeny."

---

* As is jalapeño—though according to psychologist Paul Rozin, Mexican dogs, unlike American dogs, enjoy a little heat. Rozin's work suggests animals have cultural food preferences too. Rozin was not the first academic to feed ethnic cuisine to research animals. In "The Effect of a Native Mexican Diet on Learning and Reasoning in White Rats," subjects were served chili con carne, boiled pinto beans, and black coffee. Their scores at maze-solving remained high, possibly because of an added impetus to find their way to a bathroom. In 1926, the Indian Research Fund Association compared rats who lived on chapatis and vegetables with rats fed a Western diet of tinned meat, white bread, jam, and tea. So repellent was the Western fare that the latter group preferred to eat their cage mates, three of them so completely that "little or nothing remained for post-mortem examination."

I must have registered surprise at the disclosure, because Theresa jumps in. "If you open up a bag and it smells really good—"

The tech shrugs. "And you're hungry . . ."

I N 1973 THE nutritional watchdog group Center for Science in the Public Interest (CSPI) published a booklet titled *Food Scorecard*, which made the claim that one-third of the canned dog food purchased in housing projects was consumed by people. Not because they'd developed a taste for it, but because they couldn't afford a more expensive meat product. (When a reporter asked where the figure had come from, CSPI founder Michael Jacobson couldn't recall, and to this day the organization has no idea.)

To my mind, the shocker was in the scores themselves. Thirty-six common American protein products were ranked by overall nutritional value. Points were awarded for vitamins, calcium, and trace minerals, and subtracted for added corn syrup and saturated fats. Jacobson—believing that poor people were eating significant amounts of pet food, and/or exercising his talent for publicity— included Alpo in the rankings. It scored 30 points, besting salami and pork sausage, fried chicken, shrimp, ham, sirloin steak, McDonald's hamburgers, peanut butter, pure-beef hotdogs, Spam, bacon, and bologna.

I mention the CSPI rankings to Nancy Rawson. We are back at AFB headquarters, with Moeller again, in a different conference room. (There are five: Dalmatian, Burmese, Greyhound, Calico, and Akita. The staff refer to them by breed, as in, "Do you want to go into Greyhound?" "Is Dalmatian free at noon?") It would seem that in terms of nutrition, there was no difference between the cheap meatball sub I ate for lunch and the Smart-

Blend the dogs were enjoying earlier. Rawson disagrees. "Your sandwich was probably less complete, nutritionally."

The top slot on the CSPI scorecard, with 172 points, is beef liver. Chicken liver and liver sausage took second and third place. A serving of liver provides half the RDA for vitamin C, three times the RDA for riboflavin, nine times the vitamin A in the average carrot, plus good amounts of vitamins B12, B6, and D, folic acid, and potassium.

What's the main ingredient in AFB's dog-food palatants?

"Liver," says Moeller. "Mixed with some other viscera. The first part that a wild animal usually eats in its kill is the liver and stomach, the GI tract." Organs in general are among the most nutritionally giving foods on Earth. A serving of lamb spleen has almost as much vitamin C as a tangerine. Beef lung has 50 percent more. Stomachs are especially valuable because of what's inside them. The predator benefits from the nutrients of the plants and grains in the guts of its prey. "Animals have evolved to survive," Rawson says. They like what's best for them. People blanch to see "fish meal" or "meat meal" on a pet-food ingredient panel, but meal—which variously includes organs, heads, skin, and bones—most closely resembles the diet of dogs and cats in the wild. Muscle meat is a grand source of protein, but comparatively little else.

Animals' taste systems are specialized for the niche they occupy in the environment. "That's driven their sensory systems down a certain path," Rawson says. This includes the animal known as us. As hunters and foragers of the dry savannah, our earliest forebears evolved a taste for important but scarce nutrients: salt and high-energy fats and sugars. On the African veldt, unlike at the American food court, fats, sugar, and salt were not easy to come by. That, in a nutshell, explains the widespread popularity of junk food. And wide spreads in general.

Like dogs, humans also need a broad range of vitamins, minerals, calcium. We're omnivores. Early man didn't throw away the most nutritious parts of a carcass. Why ever do we? In 2009, the United States exported 438,000 tons of frozen livestock organs. You could lay them end to end and make a viscera equator. Figuratively speaking, they already ring the globe. Egypt and Russia are big on livers. Mexico eats our brains and lips. Our hearts belong to the Philippines.

What happened here? Why are we so squeamish? How hard would it be to go back to our healthier origins? For answers, we head to the Canadian Arctic, last stronghold of the North American organ-meat dinner.

# 3

# *Liver and Opinions*

## WHY WE EAT WHAT WE EAT
## AND DESPISE THE REST

THE NORTHERN FOOD Tradition and Health Resource Kit contains a deck of forty-eight labeled photographs of traditional Inuit foods. Most are meat, but none are steaks. Seal Heart, one is labeled. Caribou Brain, says another. The images, life-size where possible, are printed on stiff paper and die-cut, like paper dolls that you badly want to throw some clothes on. The kit I looked through belonged to Gabriel Nirlungayuk, a community health representative from Pelly Bay, a hamlet in Canada's Nunavut territory. Like me, he was visiting Igloolik—a town on a small island near Baffin Island—to attend an Arctic athletic competition.* With him was Pelly Bay's mayor at the time, Makabe Nar-

---

* The Inuit Games. Most are indoor competitions originally designed to fit in igloos. Example: the Ear Lift: "On a signal, the competitor walks forward lifting the weight off the floor and carrying it with his ear for as far a distance as his ear will allow." For the Mouth Pull, opponents stand side by side, shoul-

62 · *Mary Roach*

tok. The three of us met by chance in the kitchen of Igloolik's sole lodgings, the Tujormivik Hotel.

Nirlungayuk's job entailed visiting classrooms to encourage young Inuit "chip-aholics and pop-aholics" to eat like their elders. As the number of Inuit who hunt has dwindled, so has the consumption of organs (and other anatomy not available for purchase at the Igloolik Co-op: tendons, blubber, blood, head).

I picked up the card labeled Caribou Kidney, Raw. "Who actually eats this?"

"I do," said Nirlungayuk. He is taller than most Inuit, with a prominent, thrusting chin that he used to indicate Nartok. "He does."

Anyone who hunts, the pair told me, eats organs. Though the Inuit (in Canada, the term is preferred over *Eskimo*) gave up their nomadic existence in the 1950s, most adult men still supplemented the family diet with hunted game, partly to save money. In 1993, when I visited, a small can of Spork, the local Spam, cost $2.69. Produce arrives by plane. A watermelon might set you back $25. Cucumbers were so expensive that the local sex educator did his condom demonstrations on a broomstick.

I asked Nartok to go through the cutouts and show me what he ate. He reached across the table to take them from me. His arms were pale to the wrist, then abruptly brown. The Arctic suntan could be mistaken, at a glance, for gloves. He peered at the cutouts through wire-rim glasses. "Caribou liver, yes. Brain. Yes, I

---

ders touching and arms around each other's necks as if they were dearest friends. Each grabs the outside corner of his opponent's mouth with his middle finger and attempts to pull him over a line drawn in the snow between them. As so often is the case in life, "strongest mouth wins."

eat brain. I eat caribou eyes, raw and cooked." Nirlungayuk looked on, nodding.

"I like this part very much." Nartok was holding a cutout labeled Caribou Bridal Veil. This is a prettier way of saying "stomach membrane." It was dawning on me that eating the whole beast was a matter not just of economics but of preference. At a community feast earlier in the week, I was offered "the best part" of an Arctic char. It was an eye, with fat and connective tissue dangling off the back like wiring on a headlamp. A cluster of old women stood by a chain-link fence digging marrow from caribou bones with the tilt-headed focus nowadays reserved for texting.

For Arctic nomads, eating organs has, historically, been a matter of survival. Even in summer, vegetation is sparse. Little beyond moss and lichen grows abundantly on the tundra. Organs are so vitamin-rich, and edible plants so scarce, that the former are classified, for purposes of Arctic health education, both as "meat" and as "fruits and vegetables." One serving from the Fruits and Vegetables Group in Nirlungayuk's materials is "1/2 cup berries or greens, or 60 to 90 grams of organ meats."

Nartok shows me an example of Arctic "greens": cutout number 13, Caribou Stomach Contents. Moss and lichen are tough to digest, unless, like caribou, you have a multichambered stomach in which to ferment them. So the Inuit let the caribou have a go at it first. I thought of Pat Moeller and what he'd said about wild dogs and other predators eating the stomachs and stomach contents of their prey first. "And wouldn't we all," he'd said, "be better off."

If we could strip away the influences of modern Western culture and media and the high-fructose, high-salt temptations of the junk-food sellers, would we all be eating like Inuit elders, instinctively gravitating to the most healthful, nutrient-diverse foods? Perhaps. It's hard to say. There is a famous study from the 1930s involving a group of orphanage babies who, at mealtimes, were presented with a smorgasbord of thirty-four whole, healthy foods. Nothing was processed or prepared beyond mincing or mashing. Among the more standard offerings—fresh fruits and vegetables, eggs, milk, chicken, beef—the researcher, Clara Davis, included liver, kidney, brains, sweetbreads, and bone marrow. The babies shunned liver and kidney (as well as all ten vegetables, haddock, and pineapple), but brains and sweetbreads did *not* turn up among the low-preference foods she listed. And the most popular item of all? Bone marrow.

A T HALF PAST ten, the sky was princess pink. There was still enough light to make out the walrus appliqués on the jacket of a young girl riding her bicycle on the gravel road through town. We were joined in the kitchen by a man named Marcel, just back from a hunting camp where a pod of narwhal had been spotted earlier in the day. The narwhal is a medium-sized whale with a single tusk protruding from its head like a birthday candle.

Marcel dropped a white plastic bag onto the table. It bounced slightly on landing. "Muktuk," Nirlungayuk said approvingly. It was a piece of narwhal skin, uncooked. Nartok waved it off. "I ate muktuk earlier. Whole lot." In the air he outlined a square the size of a hardback book.

Nirlungayuk speared a chunk on the tip of a pocketknife blade and held it out for me. My instinct was to refuse it. I'm a product of my upbringing. I grew up in New Hampshire in the 1960s, when meat meant muscle. Breast and thigh, burgers and chops. Organs were something you donated. Kidney was a shape for coffee tables. It did not occur to my people to fix innards for supper, especially raw ones. Raw outards seemed even more unthinkable.

I pulled the rubbery chunk from Nirlungayuk's knife. It was cold from the air outside and disconcertingly narwhal-colored. The taste of muktuk is hard to pin down. Mushrooms? Walnut? There was plenty of time to think about it, as it takes approximately as long to chew narwhal as it does to hunt them. I know you won't believe me, because I didn't believe Nartok, but muktuk is exquisite (and, again, healthy: as much vitamin A as in a carrot, plus a respectable amount of vitamin C).

I like chicken skin and pork rinds. Why the hesitation over muktuk? Because to a far greater extent than most of us realize, culture writes the menu. And culture doesn't take kindly to substitutions.

What Gabriel Nirlungayuk was trying to do with organs for health, the United States government tried to do for war. During World War II, the U.S. military was shipping so much meat overseas to feed troops and allies that a domestic shortage loomed. According to a 1943 *Breeder's Gazette* article, the American soldier consumed close to a pound of meat a day. Beginning that year, meat on the homefront was rationed—but only the mainstream cuts. You could have all the organ meats you wanted. The army didn't use them because they spoiled more quickly and because, as *Life* put it, "the men don't like them."

Civilians didn't like them any better. Hoping to change this, the National Research Council (NRC) hired a team of anthropologists, led by the venerable Margaret Mead, to study American food habits. How do people decide what's good to eat, and how do you go about changing their minds? Studies were under-taken, recommendations drafted, reports published—including Mead's 1943 opus "The Problem of Changing Food Habits: Report of the Committee on Food Habits," and if ever a case were to be made for word-rationing, there it was.

The first order of business was to come up with a euphemism. People were unlikely to warm to a dinner of "offal" or "glandular meats," as organs were called in the industry.* "Tidbits" turned up here and there—as in *Life*'s poetic "Plentiful are these meats called 'tidbits'"—but "variety meats" was the standout winner. It had a satisfactorily vague and cheery air, calling to mind both protein and primetime programming with dance numbers and spangly getups. In the same vein—ew! Sorry. *Similarly*, meal planners and chefs were encouraged "to give special attention to the naming" of new organ-meat entrées. A little French was thought to help things go down easier. A 1944 *Hotel Management* article included recipes for "Brains à la King" and "Beef Tongue Piquant."

Another strategy was to target kids. "The human infant enters the world without information about what is edible and what is not," wrote psychologist Paul Rozin, who studied disgust for many years at the University of Pennsylvania. Until kids are around two, you can get them to try pretty much anything, and Rozin did. In one memorable study, he tallied the percentage of children aged sixteen to twenty-nine months who ate or tasted

---

* Among themselves, meat professionals speak a jolly slang. "Plucks" are thoracic viscera: heart, lungs, trachea. Spleens are "melts," rumens are "paunch," and unborn calves are "slunks." I once saw a cardboard box outside a New York meat district warehouse with a crude sign taped to it: FLAPS AND TRIANGLES.

the following items presented to them on a plate: fish eggs (60 percent), dish soap (79 percent), cookies topped with ketchup (94 percent), a dead (sterilized) grasshopper (30 percent), and artfully coiled peanut butter scented with Limburger cheese and presented as "dog-doo" (55 percent). The lowest-ranked item, at 15 percent acceptance, was a human hair.*

By the time children are ten years old, generally speaking, they've learned to eat like the people around them. Once food prejudices are set, it is no simple task to dissolve them. In a separate study, Rozin presented sixty-eight American college students with a grasshopper snack, this time a commercially prepared honey-covered variety sold in Japan. Only 12 percent were willing to try one.

So the NRC tried to get elementary schools involved. Home economists were urged to approach teachers and lunch planners. "Let's do more than say 'How do you do' to variety meats; let's make friends with them!" chirps Jessie Alice Cline in the February 1943 *Practical Home Economics.* The War Food Administration pulled together a *Food Conservation Education* brochure with suggested variety-meat essay themes ("My Adventures in Eating New Foods"). Perhaps sensing the futility of trying to get ten-year-olds to embrace brains and hearts, the administration focused mainly on not wasting food. One suggested student activity took the form of "a public display of wasted edible food actually found in

---

* The children were wise to be wary. Compulsive hair-eaters wind up with trichobezoars—human hairballs. The biggest ones extend from stomach into intestine and look like otters or big hairy turds and require removal by stunned surgeons who run for their cameras and publish the pictures in medical journal articles about "Rapunzel syndrome." Bonus points for reading this footnote on April 27, National Hairball Awareness Day.

the garbage dump," which does more than say "How do you do" to a long night of parental phone calls.

The other problem with classroom-based efforts to change eating habits was that children don't decide what's for dinner. Mead and her team soon realized they had to get to the person they called the "gatekeeper"—Mom. Nirlungayuk reached a similar conclusion. I tracked him down, seventeen years later, and asked him what the outcome of his country-foods campaign had been. "It didn't really work," he said, from his office in the Nunavut department of wildlife and environment. "Kids eat what parents make for them. That's one thing I didn't do is go to the parents."

Even that can flop. Mead's colleague Kurt Lewin, as part of the NRC research, gave a series of lectures to homemakers on the nutritional benefits of organ meats, ending with a plea for patriotic cooperation.* Based on follow-up interviews, just 10 percent of the women who'd attended had gone home and prepared a new organ meat for the family. Discussion groups were more effective than lectures, but guilt worked best of all. "They said to the women, 'A lot of people are making a lot of sacrifices in this war,'" says Brian Wansink, author of "Changing Eating Habits on the Home Front." "'You can do your part by trying organ meats.' All of a sudden, it was like, 'Well, I don't want to be the only person not doing my part.'"

Also effective: pledges. Though it now seems difficult to picture it, Wansink says government anthropologists had PTA members stand up and recite, "I will prepare organ meats at least ____ times in the coming two weeks." "The act of making a public commit-

---

* Meat and patriotism do not fit naturally together, and sloganeering proved a challenge. The motto "Food Fights for Freedom" would seem to inspire cafeteria mayhem more than personal sacrifice.

ment," said Wansink, "was powerful, powerful, powerful." A little context here: The 1940s was the heyday of pledges and oaths.* In Boy Scout halls, homerooms, and Elks lodges, people were accustomed to signing on the dotted line or standing and reciting, one hand raised. Even the Clean Plate Club—dreamed up by a navy commander in 1942—had an oath: "I, ____, being a member in good standing . . . , hereby agree that I will finish all the food on my plate . . . and continue to do so until Uncle Sam has licked the Japs and Hitler"—like, presumably, a plate.

To open people's minds to a new food, you sometimes just have to get them to open their mouths. Research has shown that if people try something enough times, they'll probably grow to like it. In a wartime survey conducted by a team of food-habits researchers, only 14 percent of the students at a women's college said they liked evaporated milk. After serving it to the students sixteen times over the course of a month, the researchers asked again. Now 51 percent liked it. As Kurt Lewin put it, "People like what they eat, rather than eat what they like."

The phenomenon starts early. Breast milk and amniotic fluid carry the flavors of the mother's foods, and studies consistently

---

* Pledge madness peaked in 1942. The June issue of *Practical Home Economics* reprinted a twenty-item Alhambra, California, Student Council antiwaste pledge that included a promise to "drive carefully to conserve rubber" and another to "get to class on time to save paper on tardy slips." Perhaps more dire than the shortages in metal, meat, paper, and rubber was the "boy shortage" mentioned in an advice column on the same page. "Unless you do something about it, this means empty hours galore!" Luckily, the magazine had some suggestions. An out-of-fashion bouclé suit could be "unraveled, washed, tinted and reknitted" to make baby clothes. Still bored? "Take two worn rayon dresses and combine them to make one Sunday-best that looks brand new"— and fits like a dream if you are a giant insect or person with four arms.

show that babies grow up to be more accepting of flavors they've sampled while in the womb and while breastfeeding. (Babies swallow several ounces of amniotic fluid a day.) Julie Mennella and Gary Beauchamp of the Monell Chemical Senses Center have done a great deal of work in this area, even recruiting sensory panelists to sniff* amniotic fluid (withdrawn during amniocentesis) and breast milk from women who had and those who hadn't swallowed a garlic oil capsule. Panelists agreed: the garlic-eaters' samples smelled like garlic. (The babies didn't appear to mind. On the contrary, the Monell team wrote, "Infants . . . sucked more when the milk smelled like garlic.")

As a food marketing consultant, Brian Wansink was involved in efforts to increase global consumption of soy products. Whether one succeeds at such an undertaking, he found, depends a great deal on the culture whose diet you seek to change. Family-oriented countries where eating and cooking are firmly bound by tradition—Wansink gives the examples of China, Colombia, Japan, and India—are harder to infiltrate. Cultures like the United States and Russia, where there's less cultural pressure to follow tradition and more emphasis on the individual, are a better bet.

Price matters too, though not always how you think it would.

---

* They are to be excused for not tasting it too. Amniotic fluid contains fetal urine (from swallowed amniotic fluid) and occasionally meconium: baby's first feces, composed of mucus, bile, epithelial cells, shed fetal hair, and other amniotic detritus. The *Wikipedia* entry helpfully contrasts the tarry, olive-brown smear of meconium—photographed in a tiny disposable diaper—with the similarly posed yellowish excretion of a breast-fed newborn, both with an option for viewing in the magnified resolution of 1,280 × 528 pixels.

Saving money can be part of the problem. The well-known, long-standing cheapness of offal, Mead wrote, condemned it to the wordy category "edible for human beings but not by own kind of human being." Eating organs, in 1943, could degrade one's social standing. Americans preferred bland preparations of muscle meat partly because for as long as they could recall, that's what the upper class ate.

So powerful are race- and status-based disgusts that explorers have starved to death rather than eat like the locals. British polar exploration suffered heavily for its mealtime snobbery. "The British believed that Eskimo food . . . was beneath a British sailor and certainly unthinkable for a British officer," wrote Robert Feeney in *Polar Journeys: The Role of Food and Nutrition in Early Exploration.* Members of the 1860 Burke and Wills expedition to cross Australia fell prey to scurvy or starved in part because they refused to eat what the indigenous Australians ate. Bugong-moth abdomen and witchetty grub may sound revolting, but they have as much scurvy-battling vitamin C as the same size serving of cooked spinach, with the additional benefits of potassium, calcium, and zinc.

Of all the so-called variety meats, none presents a steeper challenge to the food persuader than the reproductive organs. Good luck to Deanna Pucciarelli, the woman who seeks to introduce mainstream America to the culinary joys of pig balls. "I am indeed working on a project on pork testicles," said Pucciarelli, director of the Hospitality and Food Management Program at— fill my heart with joy!—Ball State University. Because she was bound by a confidentiality agreement, Pucciarelli could not tell me who would be serving them or why or what form they would take. Setting aside alleged fertility enhancers and novelty dare items (for example, "Rocky Mountain oysters"), the reproductive equipment seem to have managed to stay off dinner plates worldwide. Neither I nor Janet Riley, spokesperson for the American Meat Institute, could come up with a contemporary culture that regularly partakes of ovaries, uterus, penis, or vagina simply as something good to eat.

Historically, there was ancient Rome. Bruce Kraig, president of the Culinary Historians of Chicago, passed along a recipe from *Apicius*, for sow uterus sausage. For a cookbook, *Apicius* has a markedly gladiatorial style. "Remove the entrails by the throat before the carcass hardens immediately after killing," begins one recipe. Where a modern recipe might direct one to "salt to taste," the uterus recipe says to "add cooked brains, as much as is needed." Sleeter Bull,* the author of the 1951 book *Meat for the Table*, claims the ancient Greeks had a taste for udders. Very specifically, "the udders of a sow just after she had farrowed but before she had suckled her pigs." That is either the cruelest culinary practice in history or so much Sleeter bull.

I would wager that if you look hard enough, you will find a welcoming mouth for any safe source of nourishment, no matter how unpleasant it may strike you. "If we consider the wide range of foods eaten by all human groups on earth, one must . . . question whether any edible material that provides nourishment with no ill effects can be considered inherently disgusting," writes the food scientist Anthony Blake. "If presented at a sufficiently early age with positive reinforcement from the childcarer, it would

---

* Bull was chief of the University of Illinois Meats Division and founding patron of the Sleeter Bull Undergraduate Meats Award. Along with meat scholarship, Bull supported and served as grand registrar of the Alpha Gamma Rho fraternity, where they knew a thing or two about undergraduate meats.

become an accepted part of the diet." As an example, Blake mentions a Sudanese condiment made from fermented cow urine and used as a flavor enhancer "very much in the way soy sauce is used in other parts of the world."

The comparison was especially apt in the summer of 2005, when a small-scale Chinese operation was caught using human hair instead of soy to make cheap ersatz soy sauce. Our hair is as much as 14 percent L-cysteine, an amino acid commonly used to make meat flavorings and to elasticize dough in commercial baking. How commonly? Enough to merit debate among scholars of Jewish dietary law, or kashrut. "Human hair, while not particularly appetizing, is Kosher," states Rabbi Zushe Blech, the author of *Kosher Food Production*, on Kashrut.com "There is no 'guck' factor," Blech maintained, in an e-mail. Dissolving hair in hydrochloric acid, which creates the L-cysteine, renders it unrecognizable and sterile. The rabbis' primary concern had not to do with hygiene but with idol worship. "It seems that women would grow a full head of hair and then shave it off and offer it to the idol," wrote Blech. Shrine attendants in India have been known to surreptitiously collect the hair and sell it to wigmakers, and some in kashrut circles worried they might also be selling it to L-cysteine* producers. This proved not to be the case. "The hair used in the process comes exclusively from local barber shops," Blech assures us. *Phew.*

---

* The other common source of L-cysteine is feathers. Blech has a theory that this might explain the medicinal value of chicken soup, a recipe for which can be found in the Gemorah (shabbos 145b) portion of the Talmud. L-cysteine, he says, is similar to the mucus-thinning drug acetylcysteine. And it is found, albeit in lesser amounts, in birds' skin. "Chicken soup and its L-cysteine," Blech said merrily, may indeed be "just what the doctor ordered."

• • •

THE MOST EFFECTIVE agent of dietary change is the adulated eater—the king who embraces whelks, the revolutionary hero with a passion for skewered hearts. "Normally disgusting substances or objects that are associated with admired . . . persons cease to be disgusting and may become pleasant," writes Paul Rozin. For organ meats today, that person has been taking the form of celebrity chefs at high-profile eateries, such as Los Angeles's Animal and London's St. John, and on Food Network programs. On the *Iron Chef* episode "Battle Offal," judges swooned over raw heart tartar, lamb's liver truffles, tripe, sweetbreads, and gizzard. If things go as they usually go, hearts and sweetbreads might start to show up on home dinner tables in five or ten years.

Time and again, AFB's Pat Moeller has watched the progression with ethnic cuisines: from upscale restaurant to local eatery to dinner table to supermarket freezer section. "It starts as an appetizer typically. That's low risk. Then it migrates to an entrée dish. Then it becomes a food that you can buy and take home and fix for your family."

With organ meats, where the prep may include, say, "removal of membrane," the last phase will be slow-going. Unlike filets and stewing meats, organs look like what they are: body parts. That's another reason we resist them. "Organs," says Rozin, "remind us of what we have in common with animals." In the same way a corpse spawns thoughts of mortality, tongues and tripe send an unwelcome message: you too are an organism, a chewing, digesting sack of guts.

To eat liver, knowing that you, too, have a liver, brushes up against the cannibalism taboo. The closer we are to a species, emotionally or phylogenetically, the more potent our horror at the prospect of tucking in, the more butchery feels like murder. Pets and primates, wrote Mead, come under the category "unthinkable to eat." The same cultures that eat monkey meat have traditionally drawn the line at apes.

The Inuit, at the time I visited Igloolik, had no tradition of keeping animals as companions. A sled dog was more or less a piece of equipment. When I told Makabe Nartok that I had a cat, he asked, "What do you use it for?" In America, pets are family, never fare. That feeling held fast even during the years of World War II rationing, when horse or rabbit—delicacies right over the pond in France—might, you'd think, have seemed preferable to organs. In the 1943 opinion piece "Jackrabbit Should Be Used to Ease Meat Shortage," Kansas City scientist B. Ashton Keith bemoans the "wasted meat resource" of jackrabbit carcasses that were being left for coyotes and crows after being killed by ranchers in "great drives that slaughter thousands." (Most of these seemingly collected by Keith's mother: "Some of the pleasantest recollections of my boyhood are of fried jackrabbit, baked jackrabbit, jackrabbit stew, and jackrabbit pie.")

SELF-MADE "NUTRITIONAL ECONOMIST" Horace Fletcher espoused a singular approach to getting Americans through a wartime meat shortage without resorting to rationing, or jackrabbits. What Fletcher proposed was a simple if burdensome adjustment to the human machinery.

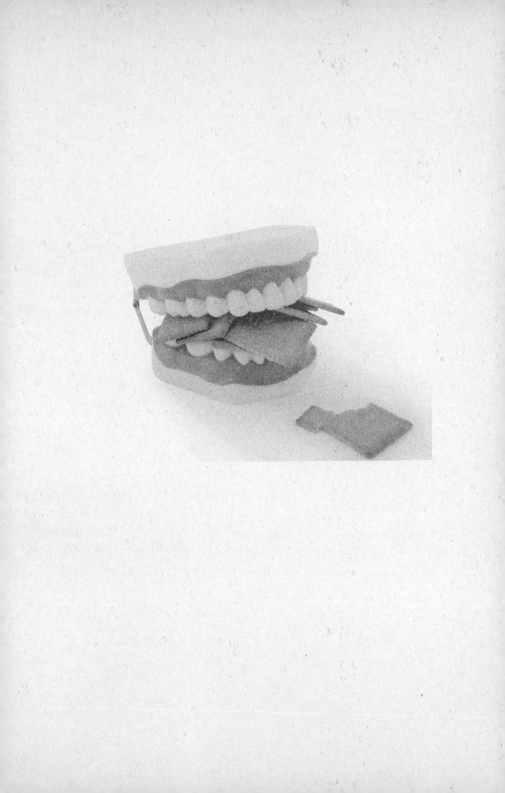

# *4*

# *The Longest Meal*

## CAN THOROUGH CHEWING
## LOWER THE NATIONAL DEBT?

*T*HE HORACE FLETCHER papers reside in a single cardboard box of a size that would hold a lightweight cardigan. The
self-dubbed economic nutritionist did not attend Harvard,\* but
it is Harvard that came to own his letters, now stored in some dim
recess of the Houghton Library. It was a spring day in May when
I visited them. Outside the open windows, a run-through of
commencement was under way, speeches being delivered before a
plain of empty chairs. I recall feeling relieved by the compactness
of the collection, for it appeared it could be gone through in a

---

\* He did, however, leave the residue of his estate to Harvard, part of which
went toward funding the Horace Fletcher Prize. This was to be awarded each
year for "the best thesis on the subject 'Special Uses of Circumvallate Papilli
and the Saliva of the Mouth in Regulating Physiological Economy in Nutrition.'" Harvard's Prize Office has no record of anyone applying for, much less
winning, the prize.

couple of hours, leaving time to enjoy the warm, chlorophyll-brightened Cambridge afternoon.

The box was deceptive. Fletcher had typed on the thinnest of onion-skin typing paper. As the years wore on, his margins grew smaller and smaller, often disappearing entirely. Fletcher was an efficiency buff, and his obsession appears to have carried over to his habits as a letter writer. Just as he believed in extracting the most nourishment possible from a mouthful of food, he sought to extract the maximum use from each sheet of stationery. Around 1913, he switched from double- to single-spacing and began typing on both sides of the page. Because the paper was thin to the point of transparency, Fletcher's words bled through, causing some of the missives, though typed, to be practically illegible.

What I am getting at is that there is a point at which efficiency crosses over into lunacy, and the savings in money or resources cease to be worthwhile in light of the price paid in other ways. Horace Fletcher danced around that point his whole career. What amazes me is the degree to which he was taken seriously.

Fletcher was the instigator of a fad for extremely thorough chewing. We are not talking about British Prime Minister William Gladstone's thirty-two chews per bite. We are talking about this: "One-fifth of an ounce of the midway section of the young garden onion, sometimes called 'challot,' has required seven hundred and twenty-two mastications before disappearing through involuntary swallowing." (More on chewing and the "oral device" in chapter 7.)

Fletcher in the flesh did not, by most accounts, appear to be the crackpot that that sentence suggests. He is described as cheerful and charming, a bon vivant who liked to dress in cream-colored suits that set off his tan and matched his snowy hair. He believed in physical fitness, clean living, good manners, fine food.

Fletcher's well-lubed charm and connections served him well.

Generals and presidents took up "Fletcherizing," as did Henry James, Franz Kafka, the inevitable Sir Arthur Conan Doyle. In 1912, around the fad's peak, Oklahoma Senator Robert L. Owen penned a proclamation—a draft of which resides among the Fletcher papers—urging the formation of a National Department of Health based on the principles of the Fletcher system. Senator Owen declared excessive chewing a "national asset" worthy of compulsory teaching in schools. Not long after, Fletcher snagged a post on Herbert Hoover's World War I Commission for Relief in Belgium.

It was not mere charisma that landed him there. Fletcherism held a good deal of intuitive appeal. Fletcher believed—decided, really—that by chewing each mouthful of food until it liquefies, the eater could absorb more or less double the amount of vitamins and other nutrients. "Half the food commonly consumed is sufficient for man," he stated in a letter in 1901. Not only was this economical—Fletcher estimated that the United States could save half a million dollars a day by Fletcherizing—it was healthier, or so he maintained. By delivering heaps of poorly chewed food to the intestine, Fletcher wrote, we overtax the gut and pollute the cells with the by-products of "putrid bacterial decomposition." While other feces-fearers of the day advocated enemas to speed food through the putrefaction zone (and more on this in chapter 14), Fletcher advised delivering less material.

Practitioners of Fletcher's hyperefficient chewing regimen, he wrote, should produce one-tenth the bodily waste considered normal in the health and hygiene texts of his day. And of a superior quality—as demonstrated by an unnamed "literary test subject" who, in July 1903, while living in a hotel in Washington, D.C., subsisted on a glass of milk and four Fletcherized corn muffins a day. It was a maximally efficient scenario. At the end of eight days, he had produced sixty-four thousand words, and just one BM.

"Squatting upon the floor of the room, without any perceptible effort he passed into the hollow of his hand the contents of the rectum . . . ," wrote the anonymous writer's physician in a letter printed in one of Fletcher's books. "The excreta were in the form of nearly round balls," and left no stain on the hand. "There was no more odour to it than there is to a hot biscuit." So impressive, so clean, was the man's residue that his physician was inspired to set it aside as a model to aspire to. Fletcher adds in a footnote that "similar [dried] specimens have been kept for five years without change," hopefully at a safe distance from the biscuits.

At one chew per second, the Fletcherizing of a single bite of shallot would take more than ten minutes. Supper conversation presented a challenge. "Horace Fletcher came for a quiet dinner, sufficiently chewed," wrote the financier William Forbes in his journal from 1906. Woe befall the non-Fletcherizer forced to endure what historian Margaret Barnett called "the tense and awful silence which . . . accompanies their excruciating tortures of mastication." Nutrition faddist John Harvey Kellogg, whose sanatorium briefly embraced Fletcherism,* tried to reenliven mealtimes by hiring a quartette to sing "The Chewing Song,"† an original Kellogg composition, while diners grimly toiled. I

---

* The two parted ways over feces. Kellogg's healthful ideal was four loose logs a day; Fletcher's was a few dry balls once a week. It got personal. "His tongue was heavily coated and his breath was highly malodorous," sniped Kellogg.

† I managed to track down only one stanza. It was enough. "I choose to chew / because I wish to do / the sort of thing that Nature had in view / Before bad cooks invented sav'ry stew / When the only way to eat was to chew, chew, chew."

searched in vain for film footage, but Barnett was probably correct in assuming that "Fletcherites at table were not an attractive sight." Franz Kafka's father, she reports, "hid behind a newspaper at dinnertime to avoid watching the writer Fletcherize."

How did this unsightly and extreme practice come to be taken seriously? Fletcher, an assiduous networker and general gadabout, began by getting the scientists on his side. Though he had no background in medicine or physiology, he collected friends who did. While living in a hotel in Venice in 1900, Fletcher befriended the hotel doctor, Ernest van Someren. Originally more interested in Fletcher's stepdaughter than in his theories, van Someren was eventually won over (or worn down—Fletcher's letters, though gaily phrased,* amount to lengthy harangues). Van Someren gussied up Fletcher's theories with invented medical jargon like the "secondary reflex of deglutition."

As only a hotel doctor has time for, van Someren set to work gathering the data both men knew Fletcher would need to gain approval in scientific circles. Fletcher had experimented on himself, but these efforts were unlikely to convince the research community. He had simply weighed and recorded each day's bodily input and output for both himself and "my man Carl," over the course of a bicycle trip through France. As Fletcher described the scenario, in a letter to one of his benefactors in 1900, Carl was "a young Tyrolean . . . in national costume" hired to carry the scale and "wheel my bicycle up the grades and be generally useful."

Van Someren presented a paper at a meeting of the British Medical Association in 1901, and again at the International Congress of Physiology. Skeptical but intrigued, well-placed scientists

---

* "Vesuvius is puking lava at an alarming rate."

at London's Royal Society and at Cambridge University, and Yale's Russell Chittenden* undertook follow-up studies, with mixed conclusions. In 1904, thirteen lads of the Hospital Corps Detachment of the U.S. Army were taken away from their nursing duties for six months to serve as guinea pigs in a test of Fletcher and Chittenden's low-calorie, low-protein, super-mastication regimen. Here there was no strapping lad in knickers and feathered felt hat to do the weighing and tidying up. The men's work began at 6:45 A.M. with an hour and a half of "duties about the quarters, such as . . . assisting in measurement of urine and faeces and transportation of the same to the laboratory; cleansing of faeces cans and urine bottles, etc."

Chittenden claimed to have evidence that the Fletcher system enabled a man to get by on two-thirds the calories and half the protein recommended by the current nutrition guidelines. Though the claims were roughly critiqued and largely dismissed by other scientists, they struck a chord with victualers: military officers and others whose jobs entailed feeding hungry hordes on limited budgets. In the United States and Europe, administrators at workhouses, prisons, and schools flirted with Fletcherism. The U.S. Army Medical Department issued formal instructions for a "Method of Attaining Economic Assimilation of Nutriment"— aka the Fletcher system. ("Masticate all solid food until it is completely liquefied," begins the familiar refrain.) In 1917, Chittenden

---

* A summary of Chittenden's project appears in the June 1903 issue of *Popular Science Monthly*, on the same page as an account of the Havre Two-Legged Horse, a foal born without forelegs, resembling a kangaroo "but with less to console it, since the latter has legs in front, which, while small and short are better than none at all." On a more upbeat note, the foal was "very healthy and obtains its food from a goat."

became a scientific advisor to Herbert Hoover, then the head of the U.S. Food Administration. Fletcher, living in Belgium during World War I and already chummy with the U.S. ambassador there, parlayed these two connections into his gig as an "honorary alimentary expert" for Hoover's relief commission. Together, he and Chittenden did their best to convince Hoover to make Fletcherism part of U.S. economic policy, thereby justifying a two-thirds reduction in the amount of civilian rations shipped overseas. Hoover sagely resisted.

Fletcher's true colors could occasionally be seen through the seams of his cream-colored suits. After bragging, in a 1910 letter, that a family of five could save enough money to furnish a five-room flat in fifteen months by Fletcherizing, he adds, "Of course, the furnishings must be of the simplest sort." This from a man who lived for years in a suite at the Waldorf Astoria. He summed up his policies at the end of the letter: "Expert economics coming to the assistance of ambitious unintelligence." Let them chew cake.

The nineteenth and early twentieth centuries saw a cavalcade of possibly well-meaning but probably just greedy individuals attempting to feed the poor on a shoestring budget. In the case of Jean d'Arcet Sr. and Jr., actual leather laces would have provided more nourishment than what the two proposed. In 1817, d'Arcet Jr., a chemist by trade, came up with a method for extracting gelatin from bones (and money from Parisian welfare coffers). Public hospitals and poorhouses, having swallowed the preposterous claim that two ounces of d'Arcet's gelatin was the nutritional equivalent of three-plus pounds of meat, began serving soup made with the gelatin.

So plentiful were the complaints that in 1831, physicians at an infamous Paris hospital for the poor, Hôtel-Dieu, ran an experiment comparing traditional bouillon with gelatin-based broth. The latter was "more distasteful, more putrescible, less digestible, less nutritious, and . . . moreover, it often brought on diarrhea." The French Academy of Sciences sprang into inaction, appointing a committee to look into it. The Gelatin Commission would dither for ten years before finally issuing a thumbs-down. Gelatin fed to animals, the committee reported, was found to "excite an intolerable distaste to a degree which renders starvation preferable."

Of relevant interest, an 1859 issue of *California Farmer and Journal of Useful Sciences* offers a recipe* for a nutritional extract made from Peruvian seabird guano. While its inventor, a Mr. Win. Clark of England, recommended the elixir "for all classes of society," he deemed it especially fit for "those who have much exertion and have not the means of buying meat." Mr. Clark claimed two to three tablespoonfuls was equal to two pounds of meat, with the advantage that it lends to the laborers' potatoes and peas "a very agreeable taste!"


IN 1979, A pair of Minneapolis researchers put Fletcherism to the test. They brought ten subjects to the local Veterans Administration hospital, and purchased some peanuts and jars of peanut butter. The subjects first ate a diet in which almost all the fat came from peanuts. The peanuts were then swapped out for the peanut butter—an aesthetically acceptable stand-in for excessively chewed peanuts. The subjects' "digestion ash," as Fletcher liked to call excrement, was then analyzed to see how much of the peanut fat was leaving the body unabsorbed.

---

* "Put 2 and ½ pounds of guano with 3 qts of water in an enamel stew-pan, boil for 3 or 4 hours, then let it cool. Separate the clear liquid, and about a quart of this healthy extract is obtained." Use sparingly, cautions the author, or it "will be as repugnant as pepper or vinegar."

" 'Nature will castigate those who don't masticate' may hold some truth," concluded the paper, which appeared in the October 1980 issue of the *New England Journal of Medicine*. On the whole-peanut diet, the subjects excreted 18 percent of the fat they'd consumed. When they switched to peanut butter, only 7 percent escaped in their stool.

But peanuts are hardly representative of the average food. Everyone knows—via "visual observation of stool samples," to use the *New England Journal of Medicine*'s way of saying "a glance before flushing"—that chunks of peanuts make their way through the alimentary canal undigested. Nuts are known for this. Peanuts (and corn kernels) are so uniquely and reliably hard to break down that they are used as "marker foods" in do-it-yourself tests of bowel transit time[*]—the time elapsed between consumption and dismissal. Peanuts are the food singled out for this trait by Martin Stocks, business development manager for the Model Gut, a computerized tabletop digestive tract[†] that can be hired out for absorption studies.

I had contacted Stocks to see if it might be possible to engage the Model Gut for a test of Fletcherism. It was, but would "likely run into the $10,000 to $20,000 range." Stocks's opinion was that with a few recalcitrant foods—here he singled out nuts and rare or raw meat—extensive mastication (chewing) might make a small difference in how much energy and nutrients are absorbed, but that it was "unlikely to have a dramatic effect on one's overall nutritional intake."

---

[*] The human digestive tract is like the Amtrak line from Seattle to Los Angeles: transit time is about thirty hours, and the scenery on the last leg is pretty monotonous.

[†] "It can even vomit," boasts its designer. No reply arrived in response to an e-mail asking whether and into what the Model Gut excretes.

Stocks passed my e-mail along to Model Gut senior scientist Richard Faulks. Faulks was dismissive not only of extreme chewing, but also of the related fad for blenderizing to increase the accessibility of nutrients. It's true saliva carries an enzyme that breaks down starch, but the pancreas makes this enzyme too. So any digestive slack caused by hasty chewing would be taken up in the small intestine. The human digestive tract has evolved to extract the maximum it can from the food ingested, Faulks said, and that is probably all it needs. "Nutritional science is dogged by the idea that if some is good, more is better," he said, "and this has led to the belief that we should endeavor to extract as much as possible of whichever fashionable component is in vogue. This is to ignore evolutionary biology and the imperative of survival." He pretty much ran Horace Fletcher through the Model Gut.

One thing to be said in favor of thorough chewing is that it slows an eater down. This is helpful if that particular eater is trying to shed some weight. By the time his brain registers that his stomach is full, the plodding thirty-two-chews-per-bite eater will have packed in far less food than the five-chews-per-bite wolfer. But there's *thorough* and there's *Fletcher*. Chewing each bite, say, a hundred times, Faulks said, could have the opposite effect. It would lengthen the meal so radically that the stomach could have time to empty the earliest mouthfuls into the small intestine, while the last mouthfuls are still on the table. Thereby making room for more. Fletcherizers' meals could conceivably be so interminable that by the time they finally cleared their plate and set down their napkin, they'd already be feeling peckish again.

Not to mention, the morning's half gone. "Who has time for this?" was the reaction of Jaime Aranda-Michel, a gastroenterologist with the Mayo Foundation, when I called to ask him about

Fletcherizing. "You're going to spend all day just having breakfast. You will lose your job!"

*L*ONG BEFORE DIGESTION researchers had veterans' stool and Model Guts to play with, they had Alexis St. Martin. In the early 1800s, St. Martin worked as a trapper for the American Fur Company in what is now Michigan. At age eighteen, he was accidently shot in the side. The wound healed as an open fistulated passage, the hole in his stomach fusing with the overlying holes in the muscles and skin. St. Martin's surgeon, William Beaumont, recognized the value of the unusual aperture as a literal window into the actions of the human stomach and its mysterious juices, about which, up until that point, nothing much was known.

Experiment 1 began at noon on August 1, 1825. "I introduced through the perforation, into the stomach, the following articles of diet, suspended by a silk string: . . . a piece of high seasoned à la mode beef; a piece of raw salted fat pork; a piece of raw salted lean beef; . . . a piece of stale bread; and a bunch of raw sliced cabbage; . . . the lad continuing his usual employment about the house."

On the very first day of his research career, Beaumont's work dealt a bruising blow to Fletcherism*—seventy-five years before it

---

* More recently, the digestive action of a healthy adult male obliterated everything but 28 bones (out of 131) belonging to a segmented shrew swallowed without chewing. (Debunking Fletcher wasn't the intent. The study served as a caution to archaeologists who draw conclusions about human and animal diets based on the skeletal remains of prey items.) The shrew, but not the person who ate it, was thanked in the acknowledgments, leading me to suspect that the paper's lead author, Peter Stahl, had done the deed. He confirmed this, adding that it went down with the help of "a little bit of spaghetti sauce."

was invented: "2 p.m. Found the cabbage, bread, pork, and boiled beef all cleanly digested and gone from the string." No chewing necessary.* Only the raw beef remained intact.

Beaumont carried out more than a hundred experiments on St. Martin and eventually published a book on the work, securing his place in the history of medicine. Textbooks today still make reference to Beaumont, often with hyperbolic phrasing: "the father of American physiology," "the patron saint of American physiology." From the perspective of Alexis St. Martin, there was nothing saintly or fatherly about him.

---

* The Beaumont findings were pointed out to Fletcher in a discussion that followed a lecture of his at a 1909 dental convention in Rochester, New York. "It made no practical difference whether the food was previously masticated very thoroughly, or whether the morsel . . . was introduced . . . in one solid chunk," said an audience member. Before Fletcher could reply, two more doctors chimed with opinions on this and that. By the time Fletcher spoke again, two pages farther into the transcript, the mention of Beaumont was either forgotten or conveniently ignored. At any rate, Fletcher didn't address it.

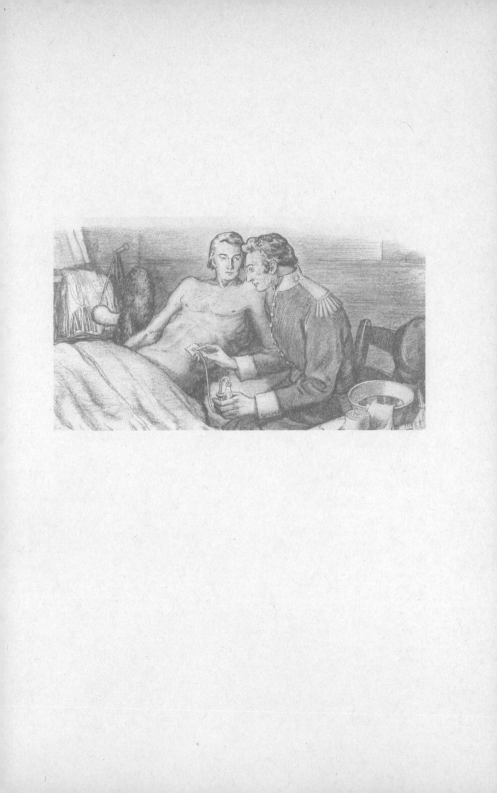

# 5

# *Hard to Stomach*

## THE ACID RELATIONSHIP OF WILLIAM BEAUMONT AND ALEXIS ST. MARTIN

**T**HREE FAMOUS ENGRAVINGS depict Alexis St. Martin in his youth. I've seen them many times, in biographies of his surgeon William Beaumont, in Beaumont's own book, in journal articles about the pair. As detailed as the artworks are, you can't tell what St. Martin looked like from examining them. All three woodcuts are of the lower portion of his left breast, and the famous hole. I could pick St. Martin's nipple out of a lineup before I could his eyes. I suppose this makes sense; Beaumont was a researcher and St. Martin his subject—more a body than a man. But the two knew each other across a span of thirty years. They lived together on and off for a decade. Over all this time, did no fondness develop? What exactly *was* their relationship? Was St. Martin mistreated, or was digesting for science the cushiest job a hard laborer could hope for?

The two first met in June 1822, at a company store on Mackinac Island, part of a trading post owned by the American Fur

Company. St. Martin was a French Canadian voyageur—an indentured trapper—hauling pelts by canoe and on foot through the woodsy landscape of the Michigan Territory. St. Martin retained little memory of the pair's historic meeting, lying, as he was, barely conscious on the floor. Someone's gun had discharged accidently, spraying a load of duck shot into St. Martin's side, and Beaumont, the army surgeon assigned to the nearby garrison, had been called down to help.

The ducks of Mackinac Island are apparently not easily taken down. "Found a portion of the Lungs as large as a turkey's egg protruding through the external wound, lacerated and burnt, and below this another protrusion resembling a portion of the Stomach, what at first view I could not believe possible to be that organ in that situation with the subject surviving, but on closer examination I found it to be actually the Stomach, with a puncture in the protruding portion large enough to receive my forefinger, and through which a portion of his food that he had taken for breakfast had come out and lodged among his apparel." Thus reads Beaumont's somewhat windy account of the injury.

Through that puncture—and in the slop of half-digested meat and bread suddenly visible in the folds of St. Martin's wool shirt—lay Beaumont's ticket to the spotlight of national renown. Italian digestion experimenters had pulled food in and out of live animal stomachs, soaked it up in sponges on strings, even regurgitated their own dinners, but St. Martin's portal presented an unprecedented opportunity to document the human juices and processes in vivo. (We will step into the stomach in earnest in chapter 8; for now, it's medicine's oddest couple that I wish to explore.)

Beaumont was thirty-seven and on the lookout for something a little glossier than the anonymous rustic toil of an assistant surgeon at a military outpost. Exactly when he realized the value of the St. Martin hole—and how assiduously he did or didn't work to close it—remain matters of conjecture. The recollection of a man named Gurdon Hubbard, the only eyewitness whose account of that morning remains, suggests the realization occurred earlier than Beaumont claimed. "I know Dr. Beaumont very well. The experiment of introducing food into the stomach through the orifice, purposefully kept open and healed with that object, was conceived by the doctor very soon after the first examination."

Beaumont denied this. In his journal, he claims to have tried "every means within my power to close the puncture of the Stomach." I imagine the truth lies midway between. Something closer to Hubbard's version would serve to explain Beaumont's puzzling dedication to a man he did not know and about whom he would have been inclined, by birthright, to care little about. St. Martin was a *mangeur du lard*—a "porkeater," the lowest class of voyageur. Yet when county funds for St. Martin's hospital care ran out, in April 1823, Beaumont moved him into his family's home. The explanation he gives in his journal was that he did so "from mere motives of charity." That I heavily doubt.

St. Martin was put to work around the house as soon as he was well enough. From the beginning, Beaumont had an eye on the fistula, more or less literally. "When he lies on the opposite side I can look directly into the cavity of the Stomach, and almost see the process of digestion," wrote Beaumont in his journal. I would love to know how the experimental protocol was first broached. St. Martin had no understanding of scientific method. He was illiterate and spoke little English. He communicated in a French Canadian patois so heavily accented that Beaumont, in his notes from the day of the shooting, transcribed "St. Martin" as "Samata." Beaumont kept diaries but neither I nor medical ethicist Jason Karlawish, who has written a fine and sleuthfully researched historical novel about the pair, could find mention of St. Martin's initial reaction to the unusual proposition.

In "Working Ethics: William Beaumont, Alexis St. Martin, and Medical Research in Antebellum America," historian Alexa Green explains the men's relationship as clearly one of master and servant." If the man wants to push a piece of mutton through your side, you let him. Other duties as assigned. (When St. Martin had healed sufficiently that the premise of providing continuing care began to seem a ruse, Beaumont provided a salary.)

For two people so firmly distanced by class and employment structure, Beaumont and St. Martin inhabited a relationship that could be oddly, intensely intimate. "On applying the tongue to the mucous coat of the stomach, in its empty, unirritated state, no acid taste can be perceived."* The one image I eventually found of Alexis St. Martin as a whole young man is in a painting by Dean Cornwell entitled *Beaumont and St. Martin*—part of the Pioneers of American Medicine series commissioned in 1938 by Wyeth Laboratories for an ad campaign. Despite the unfortunate side-parted bob that St. Martin appeared to stick with all through his adult life, the man as Cornwell rendered him is striking: broad cheekbones, vertically plunging aquiline nose, and a firmly muscled, deeply tanned chest and arms. Beaumont is dashing but dan-

---

* Using the tongue is less peculiar than it seems. Before doctors could ship patients' bodily fluids off to labs for analysis, they sometimes relied on tongue and nose for diagnostic clues. Intensely sweet urine, for instance, indicates diabetes. Pus can be distinguished from mucus, wrote Dr. Samuel Cooper in his 1823 *Dictionary of Practical Surgery,* by its "sweetish mawkish" taste and a "smell peculiar to itself." To the doctor who is still struggling with the distinction, perhaps because he has endeavored to learn surgery from a dictionary, Cooper offers this: "Pus sinks in water; mucus floats."

dified. His hair is oddly waved and piled, like something squeezed from a cake decorator's bag.

Cornwell's painting is set at Fort Crawford, in Michigan Territory, during St. Martin's second stint in Beaumont's employ, around 1830. At this stage in his digestive explorations, Beaumont had been trying to determine whether the gastric juice would work outside of the stomach, removed from the body's "vital force." (It does.) He filled vial after vial with St. Martin's secretions and dropped in all manner of foods. The cabin became a kind of gastric-juice dairy. Beaumont, in the painting, holds one end of a length of gum elastic tubing in St. Martin's stomach; the other end drips into a bottle in Beaumont's lap.

I spent a good deal of time staring at this painting, trying to parse the relationship between the two. The gulf between their stations is clear. St. Martin wears dungarees worn through at the knees. Beaumont appears in full military dress—brass-buttoned jacket with gold epaulettes, piping-trimmed breeches tucked into knee-high leather boots. "True," Cornwell seems to be saying, "it's an unsavory situation for our man St. Martin, but look, *just look,* at the splendorous man he has the honor of serving." (Presumably Cornwell took some liberties with the costuming in order to glorify his subject. Anyone who works with hydrochloric acid knows you don't wear your dress clothes in the lab.)

The emotions are hard to read. St. Martin appears neither happy nor aggrieved. He lies on his side, propped on an elbow. His posture and far-off stare suggest a man reclining by a campfire. Beaumont, admirably erect, sits in a buckskin chair by the bed. He stares into high middle distance, as though a TV set were mounted on the cabin wall. He looks like a hospital visitor who has run out of things to say. The prevailing mood of the painting is stoicism: one man enduring for the sake of science, the other for subsistence.

Given the painting's intent—the glorification of medicine (and Beaumont and Wyeth labs)—it's fair to assume the emotional content has been given a whitewash. It can't have been a hoot for either. At least once in his notes, Beaumont mentions St. Martin's "anger and impatience." The procedure was not merely tedious; it was physically unpleasant. The extraction of the gastric juices, Beaumont wrote, "is generally attended by that peculiar sensation at the pit of the stomach, termed sinking, with some degree of faintness, which renders it necessary to stop the operation."

The disrespect displayed by Beaumont and the medical establishment—evident in their correspondences about St. Martin—can't have helped. St. Martin was referred to as "the boy" well into his thirties. He was "the human test tube," "your patent digester." For the out-of-body digestion experiments, Beaumont had St. Martin hold vials of gastric juice under his arms to simulate the temperature and movements of the stomach. "Kept in the axilla and frequently agitated for one hour and half," Beaumont's notes read. If you'd never heard the term *axilla*, you'd think it was a piece of laboratory equipment, not a French Canadian's underarm. Beaumont carried out dozens of experiments that required St. Martin to hold vials this way for six, eight, eleven, even twenty-four (corn kernel!) hours. Not surprisingly, St. Martin twice quit—"absconded," as Beaumont termed it—partly to see his family in Canada, but also because he'd had enough. Only the second time did he do so in violation of a signed contract, and for this he earned Beaumont's lasting ire. In a letter to the U.S. surgeon general composed around that time, Beaumont deplores St. Martin's "villainous obstinacy and ugliness."

But Beaumont had no other fistulous stomach to turn to. Though he'd finished his experiments, he needed St. Martin to bolster his status overseas. Late in his career, he'd come to know a group

of scientists in Europe—chemists and others to whom he'd shipped*
bottles of gastric juice for analysis. (His correspondence from that
period is a mix of ghoulishness and high manners. "I thank you very
much for your Bottle of the gastric fluid." "I have . . . with peculiar
pleasure experimented upon the masticated meat . . . , as suggested
in your last letter.") Though none of these men successfully identified the various "juices," one had invited him to lecture in Europe,
with St. Martin along as a kind of human PowerPoint.

What ensued was a game of Coyote and Roadrunner that
dragged on for more than a decade. Sixty letters went back and
forth among Beaumont, St. Martin, and various contacts at the
American Fur Company who had located St. Martin and tried to
broker a return. It was a seller's market with a fevered buyer. With
each new round of communications—St. Martin holding out for
more or making excuses, though always politely and with "love to
your family"—Beaumont raised his offer: $250 a year, with an
additional $50 to relocate the wife and five children ("his live
stock," as Beaumont at one point refers to them). Perhaps a government pension and a piece of land? His final plan was to offer
St. Martin $500 a year if he'd leave his family behind, at which
point Beaumont planned to unfurl some unspecified trickery:
"When I get him alone again into my keeping I will take good
care to control him as I please." But St. Martin—*beep, beep!*—
eluded his grasp.

---

* The shipping of bodily fluids was a trying business in the 1800s. One
shipment to Europe took four months. Bottles would arrived "spilt" or "spoilt"
or both. One correspondent, taking no chances, directed Beaumont to ship
the secretions "in a Lynch & Clark's pint Congress water bottle, carefully
marked, sealed and capped with strong leather and twine, cased in tin, with
the lid soldered on."

In the end, Beaumont died first. When a colleague, years later, set out to bag the fabled stomach for study and museum display, St. Martin's survivors sent a cable that must have given pause to the telegraph operator: "Don't come for autopsy, will be killed."

$B$Y TODAY'S STANDARDS of political correctness, William Beaumont had an unattractive sense of entitlement and superiority. I don't see this as a product of flawed morals. After all, this is a man who claimed, in his diary, to be following Benjamin Franklin's "plan for attaining moral perfection." I see it, rather, as a product of nineteenth-century class structure and the larval state of medical ethics. The medical establishment of the day didn't concern itself greatly with issues of informed consent and the rights of human subjects. It wouldn't have occurred to people back then to condemn William Beaumont for exploiting a "porkeater" to advance scientific knowledge or his own career. St. Martin was compensated, they'd point out; he was never held against his will. Beaumont was judged solely on his contributions and commitment to physiology. He was, and remains, a lauded figure in the history of medicine.

More than anything else, the story of Beaumont and St. Martin is one of obsession. Here was a man who devoted his adult life and more than a thousand dollars of his own money to the study of gastric fluids. Here was a man willing, in the name of science, to taste chymified chicken from another man's stomach ("bland and sweet"). A man who became, as his biographer Jesse Myer put it, "so deeply engrossed in his subject that it was difficult for him to understand why everyone could not feel the same interest." Beaumont was crushed by the lackluster sales of his book,

*Experiments and Observations on the Gastric Juice, and the Physiology of Digestion*, in the United States and the bald disinterest from British publishers. ("I have returned *Beaumont's Experiments*, as I do not feel inclined to make an offer for it," read one rejection letter in its chilly entirety.) Among the William Beaumont Papers at Becker Medical Library are letters from the doctor to the secretary of the navy and the secretary of war, urging them to purchase a hundred copies of his book. (The navy man, a bit of a softie, bought twelve.) Beaumont had friends in high places, and he sent every single one a signed copy. Picture Martin Van Buren, then the vice president of the United States, leaning back in his magnificent leather-upholstered desk chair and opening Beaumont's book at random and reading, "At 9 o'clock A.M., I put a solid piece of rib bone, of an old hog, into a vial . . . of pure gastric juice, taken from the stomach this morning." Ambassadors, chief justices, senators, and representatives, all were forced to take time away from their weighty lives to pen thank-you notes for a book on stomach secretions. ("Truly a work of most surpassing interest." "I regret I have not yet been able to look into it with any attention.")

Obsession is a pair of blinders, and Beaumont wore his tightly. He far overstated the role of gastric acid, ignoring the digestive contributions of pepsin and of pancreatic enzymes introduced in the small intestine. As is regularly evidenced by tens of thousands of gastric reflux sufferers—their acid production pharmaceutically curtailed—humans can get by with very little gastric acid. The acid's main duty, in fact, is to kill bacteria—a fact that never occurred to Beaumont. What, for all his decades of experimenting, did he teach us? That digestion is chemical, not mechanical—but European experimenters, using animals, had shown this to be true two centuries earlier. That protein is easier to digest than vegetable matter. That gastric juices don't require the "vital forces" of the body. Not, in short, all that much.

I have on my own bookshelf a 241-page book about saliva. It is a gift from the author, Erika Silletti, and Silletti has signed it for me. She is surely as proud of her book as William Beaumont was of his, and she too endures the peculiar burden of the committed digestive scientist: the snipes and quizzical silences of people who can't understand why anyone would want to do such a thing for a living; the disappointment of parents who had looked forward to bragging about their child's career in surgery or neuroscience; the second dates that never materialize.

Dr. Silletti was delighted to hear that I wanted to visit the saliva lab. People rarely ask to visit Erika Silletti's lab. I am honestly curious about saliva, but I am also curious about obsession and its role in scientific inquiry. I think it's fair to say that some degree of obsession is a requisite for good science, and certainly for scientific breakthrough. Had I been able to spend time with William Beaumont in his lab, I imagine that my initial negative impressions of him and his work—the unorthodoxy of his methods, the seeming insensitivity to St. Martin—would have fallen away, and in their place I would have felt a measure of respect for the inventiveness and dedication at the core of what he did. I would have pitied St. Martin, not because Beaumont treated him badly, but because life had—because the circumstances of his birth afforded no opportunity to *be* William Beaumont.

Of course, there's a good chance St. Martin was happier in his simple shack with his family, "perfectly necket," than Beaumont was toiling in his labs, misunderstood by his colleagues. To each his own. Beaumont was a man for whom career came first. Like any experimenter, he was meticulous and exacting. People are messy, unpredictable things. Science you can control. Which is why St. Martin was such a bugbear for Beaumont.

Here is what William Beaumont had to say about saliva: "Its legitimate and only use, in my opinion, is to lubricate the food to facilitate the passage of the bolus through the [esophagus]." Beaumont was right about some things, but he was dead wrong about spit.

# 6

## Spit Gets a Polish

### SOMEONE OUGHT TO BOTTLE THE STUFF

 Erika Silletti studies saliva in a sunny top-floor lab in the Dutch town of Wageningen. A Gaudi poster hangs on one wall, and the windows look recently washed. The day I arrive, she wears a tailored wool skirt, short but not overly so, black leather boots, and a dove-gray cashmere sweater. If you saw a picture of Silletti in a magazine, you might make yourself feel better by assuming that the creamy skin tone and flawless symmetry of her features had been photoshopped. Only one thing fits my imagined notion of what saliva science looks like: a two-foot-tall, free-standing steel paper-towel holder with the fattest roll of paper towels I've ever seen.

I came upon Erika Silletti while roaming the abstracts of a dental conference. She later told me the presentation she gave there was met with blank looks. "They think of it as lubricating, and that's it!" She went back to her hotel room and called her boyfriend in tears.

It is safe to say that no one in this world understands and appreciates saliva like Erika Silletti.*

*H*UMANS SECRETE TWO kinds of saliva, stimulated and unstimulated, no more alike than most siblings. The prettier child is stimulated saliva. It comes from the parotid glands, between cheek and ear. When a plate of Erika Silletti's spaghetti carbonara makes your mouth water, that's stimulated saliva. It makes up 70 to 90 percent of the two to three pints of saliva each of us generates daily.

We're going to gather some now. Silletti pulls on a pair of blue latex gloves that so pleasingly complement the gray of her sweater that they look like part of the ensemble. She picks up two stoppered plastic vials. Inside each is a second, smaller vial, which contains a tightly compressed, cylindrical cotton wad. This is the Salivette saliva collection system. Silletti uncaps a Sharpie and marks an *M*, for *Mary*, on one, and an *E* on the other.

The Salivette instructions are printed in six languages. Erika Silletti, born in Italy, fluent in English, living in the Netherlands, can read three. "*Kauw dan 1 minuut lichtjes op de wattenrol.*" "*Masticate delicatamente il tampone per un minuto.*" "Gently chew the tampon for one minute." This is the simplest way to collect stimulated saliva without also collecting the food that stimulated it: you

---

* Except possibly Irwin Mandel. Mandel was the author of a hundred papers on saliva. A winner of the Salivary Research Award. The subject of a lush tribute in the *Journal of Dental Research* in 1997. The editor of the *Journal of Dental Research* in 1997. Mandel did not go so far as to write the tribute himself. That was done by B. J. Baum, P. C. Fox, and L. A. Tabak. Having three authors means no one man can be blamed for the sentence "Saliva was his vehicle and he went with the flow."

chew the collection device. This is "mechanical stimulation" (as opposed to gustatory or olfactory stimulation, which we'll come to). *Il tampone* will wick our flow, and then Silletti will place each back in its vial and put them in a centrifuge. The liquid will be spun from the cotton and flow down through an opening at the bottom of the inner vial, ending up in the outer vial.

The Salivette makes an unmistakable point: your parotid glands don't care what you chew. There is nothing remotely food-like about superabsorbent cotton, yet the parotids gamely set to work. They are your faithful servants. *Whatever you decide to eat, boss, I will help you get it down.*

Allowing you to eat is the most obvious but far from the only favor granted by saliva. Silletti removes a bottle of wine vinegar from a tote bag. With a dropper, she squirts some on my tongue. "Do you feel it? Saliva is coming in the mouth to dilute the acid." It's as though I'd taken a sip of tepid water. "The communication between the brain and mouth," says Silletti with infectious wonder. "It's *so* fast!"

Vinegar, cola, citrus juice, wine, all are in the acid range of the pH scale: from around pH 2 to 3. Anything under pH 4 will dissolve calcium phosphate, a key component of tooth enamel. The process is called demineralization. Take a drink of anything acid, and if you are paying attention, you will notice a sudden warm slosh: parotid saliva arriving like the cavalry to bring the pH back up to the safe zone. Earlier, Silletti paged through a Dutch-language textbook on saliva (*speeksel*) to show me close-up photographs of teeth belonging to dry-mouthed people—those with Sjögren's syndrome or whose salivary glands have been damaged by radiation treatments. "It's really shocking," she said, and it was: gaping brown lesions all along the gum line. "Their teeth are so soft that they cannot even eat properly."

Sugar contributes to tooth decay only indirectly. Like humans, bacteria are fond of it. "Bacteria get all crazy—party, party—they metabolize the sugar, break it down, and they release their metabolites, and *these* are acid" (though not as acid as cola or wine). In other words, sugar itself doesn't cause cavities; it's the acidic metabolites of the bacteria that feed on the sugar. As with acidic foods, saliva dilutes the acid and brings the mouth back to a neutral pH.

You may be wondering, though probably not, why newborns—who have no teeth to protect—produce excessive volumes of drool. Silletti has answers. One is simple mechanics. "They lack teeth to physically keep it in there." Your lower incisors are a sea-wall holding back the salivary tide. The other reason is the new-born's high-fat, 100 percent whole-milk diet. Baby saliva—so cute!—contains extra lipase, an enzyme that breaks down fats. (Adults have lipase mainly in their intestines.) More saliva means more lipase. As babies move on to a more varied diet, the salivary lipase tapers off.

The main digestive enzyme in stimulated saliva—everyone's, regardless of age—is amylase. In Silletti's dancing Italian accent it sounds like a liqueur or a European ingénue. Amylase breaks starches down into simple sugars that the body can use. You can taste this happening when you chew bread. A sweet taste materializes as your saliva mixes with the starch. Add a drop of saliva to a spoonful of custard, and within seconds it pours like water.

This suggests that saliva—or better yet, infant drool—could be used to pretreat food stains. Laundry detergents boast about the enzymes they contain. Are these literally digestive enzymes? I sent an e-mail to the American Cleaning Institute, which sounds like a cutting-edge research facility but is really just a trade group formerly and less spiffily known as the Soap and Detergent Association.

With no detectable appreciation for the irony of what he had written, press person Brian Sansoni referred me to a chemist named Luis Spitz. And when Dr. Spitz replied, "Sorry, I only know soap-related subjects," Sansoni—still without a trace of glee—gave me the phone number of a detergent industry consultant named Keith Grime.

When I'd composed myself sufficiently, I put in a call to Grime. The answer is yes. Higher-end detergents contain at least three digestive enzymes: amylase to break down starchy stains, protease for proteins, and lipase for greasy stains (not just edible fats but body oils like sebum). Laundry detergent is essentially a digestive tract in a box. Ditto dishwashing detergent: protease and lipase eat the food your dinner guests didn't.

Credit for the idea of using digestive enzymes for cleaning goes to chemist and Plexiglas inventor Otto Röhm. In 1913, Röhm extracted enzymes from livestock pancreases and used them to presoak dirty fabric, perhaps the clothes of the slaughterhouse staff in exchange for the pancreases; history has forgotten the details. Extracting enzymes from animal digestive tracts is costly and labor-intensive. For the first commercially produced laundry enzyme, scientists turned to a protease created by bacteria. Commercial lipase followed soon after. Here the gene was transferred to a fungus. Fungi are bigger and thus easier to deal with. You don't need a microscope to see your herd, or crop, or whatever collective noun applies to fungi.

Grime told me about an enzyme found on the forest floor that breaks down the cellulose in dead, fallen trees. When he worked at Procter & Gamble, he tried it out as a fabric softener.

That didn't work out, but the enzyme did something even better. It digested the cotton fibrils that tangle up and form pills on your sweater. (Crushingly, the anti-pilling enzyme doesn't work on wool.)

We had traveled a long distance from saliva, and I had not asked the question I'd called to ask. It was time to come in from the forest.

"If you dribble something on your shirt while you're eating," I asked Grime, "does it make sense to dab it with saliva? As a kind of natural laundry presoak?"

"That's an interesting thought."

Dr. Grime carries a Tide stain pen. He does not use his own spit.

Art conservators do. "We make cotton swabs on bamboo sticks and moisten the swab in our mouths," says Andrea Chevalier, senior paintings conservator with the Intermuseum Conservation Association. Saliva is especially helpful for fragile surfaces that solvents or water would dissolve. In 1990, a team of Portuguese conservators pitted saliva against four commonly used non-anatomical cleaning solutions. Based on its ability to clean but not damage water-gilded gold leaf and low-fired painted clay surfaces, saliva "was judged the 'best' cleaner." Denatured saliva, stripped of its enzymatic powers, was also tested and proved inferior to straight spit.

For more typical cleaning jobs, painting conservators, like laundry formulators, turn to commercially produced digestive enzymes. Protease, the protein digester, is used to dissolve washes made from egg white or hide glues. (Less enlightened conservators of yore used to spread glue made from rabbit hides onto canvases to consolidate flaking paint.) Lipase, the fat digester, is used to eat through the layers of linseed oil that eighteenth- and nineteenth-century painters applied to improve light refraction and "feed the surface" of their artworks.

Chevalier volunteered that some conservators' saliva cleans noticeably better than others', and that this occasionally prompted speculation about how many martinis these individuals were having at lunch. In reality, there are naturally large individual differences in the chemical makeup of people's saliva.

And in people's flow rates. Silletti and I, for instance, chewed our cotton wads for the same amount of time. I produced .78 milliliters of stimulated saliva; she produced 1.4. She tried to reassure me. "It doesn't say anything about how good you are or how good I am with saliva."

"Erika, I'm a dried-up husk."

"Don't say that, Mary."

Silletti excuses herself. "I want to go get some ice. The reason is that even after one minute, this will start to smell very bad."*

While she is out, I will take this opportunity to share with you the extremely surprising findings on the topic of olfactory stimulation of saliva. The notion that food smells make your mouth water is, science says, erroneous. Science has said this over and over, most recently in 1991, at King's College London. Ten subjects donned plastic odor-delivering face masks and nickel-sized

---

* I can vouch for this. I once toured the refrigerator at Hill Top Research, where odor judges test the efficacy of deodorizing products like mouthwash and cat litter. The president at the time, Jack Wild, was looking for the malodor component of armpit smell, which I had asked to sample. He kept opening little jars, going, "Nope, that's dirty feet, no, that's fishy amines" (vaginal odor). I asked him which is the worst. "Incubated saliva," he said without hesitating. "Both Thelma and I got dry heaves." I don't recall Thelma's title. Whatever she did, she deserved a raise.

Lashley cups. (The Lashley cup, a sort of glandular beret, fits on top of the parotid and collects its secretions.) Food odors wafted into the volunteers' noses: vanilla, chocolate, peppermint, tomato, and beef. Only one smell, and in only one subject, caused a significant increase in salivation. Oddly, this subject was a vegetarian, and she was smelling beef. Upon questioning, the woman revealed that the smell had nauseated her. The salivation was the kind that precedes throwing up.

It is easy to criticize that study. Sitting in a lab with a plastic mask on your face and sniffing chemically synthesized odorants does not approximate the typical mealtime mouth-watering scenario. This does, though. In 1960, a bright-eyed, full-lipped young physiologist named Alexander Kerr fried up bacon and eggs in his lab at Harvard. He did so in front of three hungry volunteers, whose parotid flow was measured via a type II outflow recorder*— the Lashley cup having not yet been invented. Even here, no one salivated any more than he had before the cooking began. The subject identified as A.G. didn't buy it. A.G. was positive he could feel his mouth "watering profusely" in the moments before he began eating. Kerr insisted that wasn't so. He told A.G. that the feeling was an artifact caused by shifting his attention to the inside of his mouth and suddenly becoming "conscious that his mouth

---

* Less high-tech than it sounds. Subjects leaned over and spat into the machine every two minutes. A slight improvement over the earliest collection technique, circa 1935: "The subject sits with head tilted forward, allowing the saliva to run to the front of the mouth . . . and drip out between parted lips." A photo in Kerr's monograph shows a nicely dressed woman, hair bobbed, hands palm down on the table in front of her, forehead resting in a support. An enamel basin is positioned just so, to catch the drippings.

contains saliva." I have seen the data, but I too find it hard to believe Dr. Kerr.

I T'S BEEN SNOWING all morning. Wet clumps of flakes flock the trunks and branches of the trees outside the lab. Silletti joins me at the window. She holds the small glass beakers that contain our fresh-from-the-centrifuge stimulated samples.

"It's beautiful," I am saying. Silletti agrees, but I notice she isn't looking out the window. Is it possible she thinks I am referring to the contents of the beakers? I'd say that, yes, it is possible. You've never seen such clear, clean-looking spit. Stimulated saliva looks, tastes, and flows like water—it is, in fact, 99 percent water. Water with some proteins and minerals. Like water from different springs, each person's saliva contains minerals in unique proportions. (People whose saliva naturally contains a lot of salt are slightly oblivious to it in their food.)

"So somebody," I observe, "could do a taste test with various salivas."

"If somebody would like to do that, yes."

Somebody—really everybody—wouldn't. I point to the beaker labeled *E.* "What about just your own? Would you ever—"

"No, I wouldn't. Even me. Although actually, you are drinking it all the time."

"Right, so—"

"*No.*"

An intriguing double standard applies to your own saliva. As long as it stays in your mouth, it's benign, welcomed even, no more offensive than the water it tastes like. Outside your mouth, it's almost as vile and contemptible as a stranger's. As part of a study, our friend from the University of Pennsylvania psychology department, Paul Rozin, asked subjects to imagine a bowl of their favorite soup and to rate their liking of it. He then asked them to rate that same bowl of soup after they'd imagined spitting into it. Forty-nine out of fifty subjects lowered their rating. Among certain castes in India, writes Edward Harper in "Ritual Pollution as an Integrator of Caste and Religion," spitting on someone puts even the spitter "in a state of severe impurity," because it is assumed that some of his saliva has "rebounded onto him."

The saliva taboo can make life burdensome for researchers. Silletti's colleague René de Wijk did a study years ago that looked at how the salivary breakdown of starch mobilizes fats and enhances flavor. (Fat is the main carrier of flavor.) To do this study, he needed his subjects to rate the taste of custard samples with and without a drop of their saliva added. You can't just have them spit in it, he explained, because then they won't go near it. He had to collect samples of their saliva without telling them why, and then add it behind their backs, like a spiteful waitress.

The same double standard applies to all "body products," as Rozin calls them, managing to make snot and saliva sound like spa purchases. We are large, mobile vessels of the very substances we find most repulsive. Provided they stay within the boundaries of the self, we feel no disgust. They're part of the whole, the thing we cherish most.

Paul Rozin has given a lot of thought to what he calls the psychological microanatomy of the mouth: Where, precisely, is the boundary between self and nonself? If you stick your tongue out of your mouth while eating and then withdraw it, does the ensalivated food now disgust you? It does not. The border of the self extends the distance of the tongue's reach. The lips too are considered an extension of the mouth's interior, and thus are part of the self. Though culture shifts the boundaries. Among religious Brahmin Indians, writes Edward Harper, even the saliva on one's own lips is considered "extremely defiling,"* to the extent that if one "inadvertently touches his fingers to his lips, he should bathe or at the least change his clothes."

The boundaries of the self are routinely extended to include the bodily substances of those we love. I'm going to let Rozin say this: "Saliva and vaginal secretions or semen can achieve positive value among lovers, and some parents do not find their young children's body products disgusting."

I recall being told, in grade school, that Eskimos kiss by rubbing noses. Is this an example of a culture reluctant to accept the saliva of a loved one? Gabriel Nirlungayuk, my go-to man for all things Eskimo/Inuit, confirmed that the *kunik*, or nose rub, has been and remains the traditional alternative to the kiss. "Even to this day, now that my children are adults, I *kunik* them when I have been away for a long time." But never girlfriends. By the time Nirlungayuk was a teenager, kissing "whiteman's way" had caught on. Nobody seemed to have a problem with extending those boundaries. If anything, the Inuit are leaders in the field. "Sometimes when my *ingutaq*—granddaughter—is full of snot, my wife or myself will wipe it clean with our mouths and then spit it out. But we would never consider this with other kids."

---

* But nothing compared to crow droppings. According to Harper, the traditional purification ritual for the Brahmin polluted by crow feces is "a thousand and one baths." This has been rendered less onerous by the invention of the showerhead and the crafty religious loophole. "The water coming through each hole counts as a separate bath."

A similar psychology applies to breast milk. It's considered natural for a child to consume a mother's milk, or even for a lover to do so, but not a stranger. (Hence the 2010 hullaballoo over the New York City restaurateur who invited diners to try cheese made from his wife's breast milk.) So reliable is breast-milk consumption as a delimiter of intimate family that Islam recognizes a category called "breast milk son," which confers an exemption to the rules on segregation of the sexes. A man can be alone with a woman if she's immediate family or if she breast-fed him as a child.* (Sisters sometimes breast-feed each other's infants, thereby creating breast-milk relatives.) Milk is thicker than blood, or about the same consistency.

$S$ILLETTI HANDS ME a plastic cup and sets a timer. We are moving on to unstimulated saliva. This is background saliva, the kind that's always flowing, though much more slowly. A minute passes. We turn away from each other and quietly spit in our cups.

"Look at the difference, compared to stimulated." Silletti tilts her cup. "You can't pour it easily. It's so viscous. Look!" She dips the

---

* Or, as of 2007, an adult. Egyptian scholar Ezzat Attiya issued a fatwa, or religious opinion, extending breast-milk-son status to anyone a woman has "symbolically breastfed." For convenience's sake, drivers and deliverymen could, by drinking five glasses of a woman's breast milk, be permitted to spend time alone with her. In the ensuing ruckus, another scholar insisted the man would have to drink directly from the woman's breast. Which is crazier: that Saudi courts, in 2009, sentenced a woman to forty lashes and four months in prison for allowing a bread deliveryman inside her home, or the notion that she might have avoided punishment by letting him suckle from her breast? The woman was seventy-five, if that helps you with your answer.

end of a glass pipette into her sample and pulls it away. *Filament* is a nice word, Silletti's word, for the mucoid strand that trails behind.

Relatively little is known about unstimulated saliva. Partly, Silletti says, because no one wants to work with it.

"Because it's so gross?"

"Because it's harder to collect. And you can't filtrate it. It clogs the filter, like hair in the drain. And you cannot be precise, because it's so slimy."

"Right, it's gross."

Silletti tucks a strand of her glossy black hair behind her ear. "It's difficult to work with."

Unstimulated saliva's trademark ropiness is due to mucins, long chains of amino acids repeating to form vast webs. Mucins are responsible for saliva's least endearing traits—its viscosity, elasticity, stickiness.* They are also responsible for some of its more heroic attributes. Unstimulated saliva forms a protective film that clings to the surfaces of the teeth. Proteins in this film bind to calcium and phosphate and serve to remineralize the enamel. Webs of mucins trap bacteria, which are then swallowed and destroyed by stomach acids. This is good, because there are a lot of bacteria in your mouth. Every time you eat something, every time you stick your finger in your mouth, you're delivering more.

---

* But not its bubbles. Frothiness is a hallmark of proteins in general; saliva has more than a thousand kinds. Proteins bind to air. When you whip cream or beat eggs, you are exposing maximum numbers of proteins to air, which is then pulled into the liquid, forming bubbles. That disturbing white foam on the cheeks and necks of racehorses is saliva whisked by the bit. (The whisking of semen is complicated by its coagulating factor. Should you wish to know more, I direct you to the mucilaginous strands of the World Wide Web.)

Picture one of those little silver* balls that cake decorators use. Strip away the metallic coating and soften the texture. You are now picturing the amassed bacteria in one milliliter of unstimulated saliva. Silletti put our samples in the centrifuge and spun cellular from noncellular. Some of what we are looking at is shed mouth cells, but most is bacteria—about a hundred million of them. More than forty species.

Yet never in my life has a cut or sore in my bacteria-crazy mouth become infected. As much as saliva is a bacterial cesspool, it is also an antimicrobial miracle—the former necessitating the latter. As a germ killer, saliva puts mouthwash to shame.† Saliva has anti-clumping properties, which discourage bacteria from forming colonies on the teeth and gums. There are salivary proteins that retain their antimicrobial abilities even when they them-

---

* Literally. The coating is real silver. That's why the label says "For Decorative Use Only." Like everyone else, environmental lawyer Mark Pollock didn't realize you weren't supposed to eat them. In 2005, Pollock sued PastryWiz, Martha Stewart Living Omnimedia, Dean & DeLuca, and a half-dozen other purveyors of silver dragees, as they are known in the business. Pollock succeeded in getting the product off store shelves in California. Fear not, holiday bakers, silver dragees are available in abundance from online sellers, along with gold dragees, mini dragees, multicolored pastel dragees. And dragee tweezers. (With cupped ends "to easily grab individual dragees.")

† As does this: Claims made by makers of mouthwash to kill 99 percent of oral bacteria are misleading. Silletti says half the species can't be cultured in a lab; they grow only in the mouth. Or *on other bacteria.* "When you ask the companies for claim support, they will show you the statistics for the kinds they can culture." How many others there are, or what mouthwash does to them, is unknown.

selves are broken down. "And they may be even more effective than the whole protein of origin," Silletti is saying. "It's incredible!"

Saliva's antimicrobial talents explain some of the folk medicine remedies that have been making the rounds since the 1600s. One 1763 treatise advocates applying "the fasting saliva of a man or woman turn'd of seventy or eighty years of age" to syphilitic chancres of the glans penis. As with the ancient Chinese *Materia Medica* prescription of saliva "applied below arms to counteract fetid perspiration," one imagines—*hopes*—that an applicator other than the tongue was employed.

"It is a known observation among the vulgar that the saliva is efficacious in cleansing foul wounds, and cicatrizing recent ones, thus dogs by licking their wounds . . . have them heal in a very short time," wrote the eighteenth-century physician Herman Boerhaave. He was correct. Wounds that would take several weeks to heal on one's skin disappear in a week inside the mouth. In a 2008 rodent study, animals that licked their wounds healed faster than those that could not (because their salivary glands had been disconnected—a wound, alas, that even saliva cannot heal).

More than just disinfecting is going on. Rodent saliva contains nerve growth factor and skin growth factor. Human saliva contains histatins, which speed wound closure independent of their antibacterial action. Dutch researchers watched it happen in the lab. They cultured skin cells, scratched them with a tiny sterile tip, soaked them in the saliva of six different people, and clocked how quickly the wounds healed, as compared to controls. Other components of saliva render viruses—including HIV, the virus that causes AIDS—noninfective in most cases. (Colds and flus aren't spread by drinking from a sick person's glass. They're spread by touching it. One person's finger leaves virus particles on the glass; the next person's picks them up and transfers them to the respiratory tract via an eye-rub or nose-pick.)*

The average person, of course, is oblivious to all this. With no more formal criteria than the number of Hollywood monsters featuring copious, pendant drool, you can make the case that saliva remains universally upsetting. And thus maligned, even in the medical community. There has long been an assumption among emergency medical personnel that human bites are especially likely to become infected and lead to sepsis—a potentially lethal systemic infection. "Even the simplest of wounds require copious irrigation and wound toilet," warn the authors of "Managing Human Bites" in the *Journal of Emergencies, Trauma, and Shock.*

Not so fast, says rival *American Journal of Emergency Medicine.* The article title says it all: "Low Risk of Infection in Selected Human Bites Treated without Antibiotics." Only one out of the sixty-two human-bit patients who were not given antibiotics developed an infection. However, the authors excluded high-risk bites, including "fight bites" on the hands. Here it is the aggressor

---

* In 1973, inquisitive cold researchers from the University of Virginia School of Medicine investigated "the frequency of exposure of nasal . . . mucosa to contact with the finger under natural conditions"—plainly said, how frequently people pick their nose. Under the guise of jotting notes, an observer sat at the front of a hospital ampitheater during grand rounds. Over the course of seven 30- to 50-minute observation periods, a group of 124 physicians and medical students picked their collective noses twenty-nine times. Adult Sunday school students were observed to pick at a slightly lower rate, not because religious people have better manners than medical personnel, but, the researchers speculated, because their chairs were arranged in a circle. In a separate phase of the study, the researchers contaminated the picking finger of seven subjects with cold virus particles and then had them pick their nose. Two of seven came down with colds. In case you needed a reason to stop picking your nose.

who gets the "bite"—when he splits open his knuckle on another man's teeth. Fight bites* tend to get infected, but it is the fault of the knuckle as much as the saliva. Relatively little blood flow reaches the tendons and sheaths of the finger joints, so the immune system has fewer resources with which to fight back. (Ear cartilage is similarly underserved by the vascular system, so if you plan on picking a fight with Mike Tyson, do practice good wound toilet.)

Even the "deathly drool" of the Komodo dragon, the world's largest lizard, has likely been overstated. Theory holds that Komodo dragon saliva contains lethal doses of infectious bacteria, enabling the reptiles to take on prey far larger than themselves— wild boar, deer, newspaper editors. (*San Francisco Chronicle*'s Phil Bronstein spent several days on an antibiotic drip after a Komodo dragon attacked his foot during a behind-the-scenes visit at the Los Angeles Zoo in 2001 with his then wife, Sharon Stone.) Rather than having to tackle and kill their prey on the spot, the theory goes, the reptiles need only deliver a single bite and then wait around for the animal to die of sepsis. The scenario has not been documented in the wild, however. A team of researchers from the University of Texas at Arlington attempted a laboratory simulation, using mice as mock prey and, as predator, injections of bacteria from wild Komodo dragon saliva. The scientists found a high death rate among mice injected with a particular bacteria, *Pasteurella multocida*. However, Australian researchers point out that *P. multocida* is common in weakened or stressed mammals. They speculate that the dragons may have picked it up from their prey, rather than the other way around. Current thinking postu-

---

* Fear the fight bite: it can cause septic arthritis. In one study, 18 of 100 cases ended in amputation of a finger. Hopefully the middle one. In the aggressive patient, a missing middle finger may be good preventive medicine.

lates a "sophisticated combined-arsenal killing apparatus," featuring venom and anticoagulative agents that lead to shock. The latter would explain "the unusual quietness . . . of prey items." Prey item Phil Bronstein was unusually *not* quiet.* "I was pretty pissed."

Though bacteria and general ropy grossness are probably to blame for saliva's nasty reputation, it may in part be lingering fallout from the writings of Hippocrates and Galen, Western medicine's most influential early (as in, triple-digit A.D. and B.C.) thinkers. Both believed sweat and saliva to be the body's way of flushing away disease-causing impurities. Before scientists realized syphilis and malaria were caused by microorganisms, the diseases were treated by putting patients in "salivating rooms." It was the same medically quaint principle that persists today in the form of taking a steam or a sauna to "sweat out toxins." Only back then, the steam included vaporized mercury to coax more saliva from the patient. No one realized that excessive salivation is a symptom of acute mercury poisoning. The salivating room was a standard feature of hospitals in the 1700s. (As was, charmingly, the "apartment for lunatics.") Patients were left inside until they'd generated six pints of saliva—about three times the amount most people produce in a day.

Not all cultures denigrate saliva. In ancient Taoist medical teachings, stimulated saliva—"the jade juice"—was said to nourish the qi, which boosts the immune defenses and, wrote one seventh-century Taoist, "puts a man beyond the reach of calami-

---

* The zookeepers, however, got very, very quiet. "So maybe," said Bronstein in an e-mail, "the dragon spit some of its quietness spray on them." I am almost 100 percent sure that that is not a reference to Sharon Stone.

ties." Given this tradition of qi-nourishing saliva retention, why do I so often see old Chinese men spitting? Silletti points out that it's not saliva being expectorated. It's phlegm from the lungs or sinuses. They spit it out, she added, because they don't care to use handkerchiefs or Kleenex. They think it's disgusting that we collect the material in our hands.

For saliva-positive attitudes, there is no place like Greece. "Greeks spit on pretty much anything they want to protect from the evil eye or bless for good luck," says Evi Numen. Numen is the exhibitions manager at the Mütter Museum,* a collection of medical curiosities amassed by Thomas Mütter and housed today at the College of Physicians of Philadelphia. Though her job qualifies her to comment on most things bodily and disgusting, her salivary expertise derives from her upbringing. Numen is of Greek extraction. Greeks spit on babies. They spit on brides. They spit on themselves. Though no actual gob is launched. "Most people," explains Numen, "say '*ftou ftou ftou*' instead of actually spitting."

The Greeks got it from the Roman Catholics, whose priests used to baptize with spittle. The priests got it from the Gospel of Mark—the bit where Jesus heals the blind by mixing dirt with his saliva and rubbing the mud on a man's eyelids. "It's an interesting passage," former Catholic priest Tom Rastrelli told me, "because the writers of the gospels of Luke and Matthew, who used Mark as

---

* Not to be confused with the Nutter D. Marvel Museum of horsedrawn carriages or the Butter Museum, a working farm that "showcases all things butter, from various styles of butter dishes to the history of butter through the ages," perhaps turning away briefly during butter's history-making 1972 role in *Last Tango in Paris*.

their source, redacted a line." Mark had included a bit about a blind man opening his eyes and seeing what looked like trees walking around. In other words, the treatment was minimally effective. The miracle of Jesus bestowing rudimentary vision to the blind doesn't have the same ring to it, so the line was cut.

THE DUTCH, BY tradition, are a dairy-farming people. Adults drink milk with dinner. A town will have a shop devoted entirely to cheese. The national dish of the Netherlands, sighs Silletti, is *vla*: custard. I have been staying in the home of food scientist René de Wijk, the world's foremost expert on the science of semi-solids like *vla*. Upon hearing this, Silletti immediately, as though it were a matter of medical urgency, invited me over for home-cooked Italian food.

Silletti is lactose-intolerant and, as concerns Dutch cuisine, just generally intolerant. "Everything is based on milk," she says, arranging sundried tomatoes for a plate of antipasto.

Silletti's home is a twenty-minute drive from Germany, where the supermarkets sell a decent range of Italian products. She regularly travels across the border to stock up. I don't blame her. The supermarket near de Wijk's house sells things like gorte pap—buttermilk barley porridge—and Smeer'm, a kind of spreadable cheese vileness. I'd go home with a cucumber and some peanuts because I wanted something real, something with crunch, something that didn't sound like a gynecology exam. There was an entire aisle devoted to *vla*.

"The Dutch and their *vla* . . ." Silletti speaks it like a curse word. "For me it's not food. You don't need teeth *or* saliva!"

Oddly, the cluster of Wageningen-area universities and research facilities known as "Food Valley" is the home of the foremost expert on the physics of crunchy food, as well as a man who knows more about chewing than anyone else in the world. I am meeting them both tomorrow, at the Restaurant of the Future. This is a cafeteria at Wageningen University where hidden cameras allow researchers to gauge how, say, lighting affects purchasing behavior, or whether people are more likely to buy bread if you let them slice it themselves. Silletti says she won't eat there.

"Because of the cameras?"

"Because of the food."

# 7

# A Bolus of Cherries

## LIFE AT THE ORAL PROCESSING LAB

*W*HEN I TOLD people I was traveling to Food Valley, I described it as the Silicon Valley of eating: fifteen thousand scientists dedicated to improving or, depending on your sentiments about processed food, compromising the quality of our meals. At the time I made the Silicon Valley comparison, I did not expect to be served actual silicone. But here it is, a bowl of rubbery white cubes the size of salad croutons. Andries van der Bilt brought them from his lab in the brusquely named Department of Head and Neck, at the nearby University Medical Center Utrecht.

"You chew them," he says.

Van der Bilt has studied chewing for twenty-five years. If a man can be said to resemble a tooth, van der Bilt is a lower incisor, long and bony with a squared-off head and a rigid, straight-backed way of sitting. It's between meals now in the camera-rigged

Restaurant of the Future. The serving line is unstaffed, and the cash registers are locked. Outside the plate-glass windows, it's snowing again. The Dutch pedal along on their bicycles, seeming daft, or photoshopped.

The cubes are made of a trademarked product called Comfort Putty, more typically used in its unhardened form for taking dental impressions. Van der Bilt isn't a dentist, however. He is an oral physiologist. He uses the cubes to quantify "masticatory performance"—how effectively a person chews. Research subjects chew a cube fifteen times and then return it in its new, un-cube-like state to van der Bilt, who pushes it through a set of sieves to see how many bits are fine enough to pass through.

I take a cube from the bowl. Van der Bilt, the cameras, and emotion-recognition software called Noldus FaceReader watch me chew. By tracking facial movements, the software can tell if customers are happy, sad, scared, disgusted, surprised, or angry about their meal selections. FaceReader may need to add a special emotion for people who have chosen to have the Comfort Putty. If you ever, as a child, chewed on a whimsical pencil eraser in the shape of an animal, say, or a piece of fruit, then you have tasted this dish.

"I'm sorry." Van der Bilt winces. "It's quite old." As though fresh silicone might be better.

The way you chew is as unique and consistent as the way you walk or fold your shirts. There are fast chewers and slow chewers, long chewers and short chewers, right-chewed people and left-chewed people. Some of us chew straight up and down, and others chew side to side like cows. Van der Bilt told me about a study in which eighty-seven people came into a lab and chewed an identical amount of shelled peanuts. Though all had a full complement of healthy teeth, the number of chews ranged from 17 to 110. In another project, subjects chewed seven foods of widely varying textures. The best predictor of how long they chewed before swallowing wasn't any particular attribute of the food. The best predictor was simply who's chewing. Your oral processing habits are a physiological fingerprint. As with the finger kind, most of us have no idea what ours look like.* We couldn't pick our own chewing mouths out of a lineup, although it would be interesting to try.

Van der Bilt studies the neuromuscular elements of chewing. You often hear about the impressive power of the jaw muscles. In terms of pressure per single burst of activity, these are the strongest muscles we have. But it is not the jaw's power to destroy that fascinates van der Bilt; it is its nuanced ability to protect. Think of a peanut between two molars, about to be crushed. At the precise millisecond the nut succumbs, the jaw muscles sense the yielding and reflexively let up. Without that reflex, the molars would continue to hurtle recklessly toward one another, now with no intact nut between. To keep your he-man jaw muscles from smashing your precious teeth, the only set you have, the body evolved an automated braking system faster and more sophisticated than anything on a Lexus. The jaw is ever vigilant. It knows its own strength. The faster and more recklessly you close your mouth, the less force

---

* Fingerprints come in three types: loop (65 percent), whorl (30 percent), and arch (5 percent). Oral processing styles for semisolid foods come in four: simple (50 percent), taster (20 percent), manipulator (17 percent), and tonguer (13 percent). Thus the millions of variations that make you the unique and delightful custard-eater and fingerprint-leaver that you are.

the muscles are willing to apply—without your giving it a conscious thought.

You can witness the protective cutout reflex by hooking up a person's jaw muscles to an electromyograph. The instant something hard gives way, the readout of electrical activity goes briefly flat. "The silent period, they call it," van der Bilt says. It seems like a term kindergarten teachers might use, or people at a Quaker meeting. All these years, I've had it backward. Teeth and jaws are impressive not for their strength but for their sensitivity. Chew on this: Human teeth can detect a grain of sand or grit ten microns in diameter. A micron is 1/25,000 of an inch. If you shrank a Coke can until it was the diameter of a human hair, the letter *O* in the product name would be about ten microns across. "If there's some earth in your salad, for instance, you notice immediately. It warns you for the wrong things." Van der Bilt did the experiment himself. "We took some *vla . . .*" Custard! In the Netherlands, *vla* is never far from where you are. "We put some plastic grains of various sizes in it . . ."

Van der Bilt stops himself. "I don't know if you want to hear these things." He has a tentative, apologetic manner of speaking, like a man accustomed to feeling that his audience, at any moment, is about to make an excuse and get up to go. Earlier he told me that his unit at Utrecht is slated to close when he retires, in a year. "There isn't," he said, "enough interest."

I think it may be something else.

THE STUDY OF oral processing is not just about teeth. It's about the entire "oral device": teeth, tongue, lips, cheeks, saliva, all working together toward a singular unpicturesque goal: bolus formation. The word *bolus* has many applications, but we are speaking of this one: a mass of chewed, saliva-moistened food particles. Food that is in—as one researcher put it, sounding like a license plate—"the swallowable state."*

I don't think the scientists are uninterested. I think they may be disgusted. This is a job where on any given day, you may find yourself documenting "intraoral bolus rolling" or shooting magnified close-ups of "retained custard" with the Wageningen University tongue-camera. Should you need to employ, say, the Lucas formula for bolus cohesiveness, you will need to figure out the viscosity and surface tension of the moistening saliva as well as the average radius of the chewed food particles and the average distance between them. To do that, you'll need a bolus. You'll need to stop your subject on the brink of swallowing and have him, like a Siamese with a hairball, relinquish the mass. If the bolus in question is a semisolid—yogurt and *vla* are not chewed, but they are "orally manipulated" and mixed with saliva—the work is yet less beautiful. As evidenced by this caption in a textbook chapter by my host René de Wijk: "Figure 2.2. Photographs of spat-out custard to which a . . . drop of black dye has been added."

Humans, even physiologists, don't like to think about food once they've begun to process it. The same chanterelle and Gorgonzola galette that had the guests swooning is, after two seconds in the mouth, an object of universal revulsion. No one knew this more intimately than Tom Little, an Irish American laborer who ate his meals by chewing food and spitting it into a funnel that fed into his stomach. When he was nine years old, in 1895, Tom swal-

* I nominate Rhode Island.

lowed a draught of clam chowder without letting it cool. The burn healed with strictures that fused the walls of his esophagus. Surgeons created a fistulous opening to his stomach so he could eat— or "feed," as Tom now referred to the act of taking in sustenance. It was an undiminishing source of embarrassment. (Interestingly, his doctor noted in a book about the case, Tom "blushed both in his face and his gastric mucosa.") He told no one, and took his meals alone or with his mother. When he finally married, it was to an older woman for whom he felt little attraction. He chose her, he told his doctor, because "she doesn't mind the way I feed."

In the bulimic community, the weight-loss strategy known as "chewing and spitting" (or CHSP) is by far the least popular. Only 8 percent of bulimic patients seen at the Eating Disorders Clinic at the University of Minnesota reported having engaged in CHSP more than three times a week—usually resorting to it only if they were unable to make themselves vomit, or because regurgitated stomach acid was damaging their teeth or esophagus. Rarely would the study's author, Jim Mitchell, encounter a patient "whose sole problem is chewing and spitting."

Of all the unflattering and untrue stories printed in the tabloids about Elton John over the years, this one drove him to sue: "Rock star Elton John's weight has plunged . . . thanks to a bizarre new habit of eating food then spitting it out." The article, which ran in London's *Sunday Mirror* in 1992, described him at a holiday party at his manager's home, spitting chewed shrimp into a napkin, commenting gaily, "'I love food, . . . but what's the point of swallowing it, you can't taste it as it goes down your throat.'" The editors admitted to having fabricated the story but didn't feel John had been defamed. The jury disagreed, awarding the singer £350,000—about $570,000—in damages.

Disgust and shame don't fully account for the unpopularity of CHSP. This does: chewing without swallowing is not eating. It doesn't scratch the itch. *That's* the point of swallowing it, made-up Elton: satisfaction. As regards eating, Mitchell told me, there's an imaginary line at the esophagus. "Everything happening above the neck—smelling, tasting, seeing—drives you toward eating, and everything below drives you toward stopping." Chewing causes saliva to be secreted, which dissolves the food and brings more of it in contact with the taste buds. Taste receptors recognize salts, sugars, fats, the things bodies need to thrive, and impel us to stock up. As the stomach fills and satiety grows, the head pipes down. Presently the plate is pushed away. When you chew food without swallowing it, the line at the neck is never crossed. The head keeps up its clamor.

Which brings us to another reason the incidence of CHSP is so low. It's expensive. Some of the women Mitchell interviewed would catch and release several dozen doughnuts at a go, flushing twenty-plus dollars down the toilet.

JIANSHE CHEN CAN tell you the flow speed of a high-viscosity bolus.* He knows the shear strength of a ricotta-cheese bolus, the deformability of Nutella, the minimum number of chews required to ready a McVitie's Digestive biscuit for the swallow (eight). On the Internet I found a copy of Chen's PowerPoint on the "dynamics of bolus formation and swallowing," so I too know these things. What I don't know is the point of it all. Chen made the

* Assuming equal terrain and baggage count, about as fast as a tortoise—.22 miles per hour.

mistake of putting his University of Leeds e-mail address on the website.

He wrote back right away. You get the sense oral processing experts are not, generally speaking, besieged by media inquiries. The aim of the work, he said, is to "provide guidance on how to formulate foods for safe eating by disadvantaged consumers." Bolus formation and swallowing depend on a highly coordinated sequence of neuromuscular events and reflexes. Disable any one of these—via stroke, degenerative neurological condition, tumor irradiation—and the seamless, moist ballet begins to fall apart. The umbrella term is dysphagia (from the Greek for "disordered eating," which may or may not explain flaming Greek cheese appetizers).

Most of the time, while you're just breathing and not swallowing, the larynx (voice box) blocks the entrance to the esophagus (food tube). When a mouthful of food or drink is ready to be swallowed, the larynx has to rise out of the way, both to yield access to the esophagus and to close off the windpipe and prevent the food from being inhaled. To allow this to happen, the bolus is held momentarily at the back of the tongue, a sort of anatomical metering light. If, as a result of dysphagia, the larynx doesn't move quickly enough, the food can head down the windpipe instead. This is, obviously, a choking hazard. More sinisterly, inhaled food and drink can deliver a troublesome load of bacteria. Infection can set in and progress to pneumonia.

A less lethal and more entertaining swallowing misstep is nasal regurgitation. Here the soft palate—home turf of the uvula,* that queer little oral stalactite—fails to seal the opening to the nasal cavity. This leaves milk, say, or chewed peas in peril of

---

* Its full medical name, and my pen name should I ever branch out and write romance novels, is palatine uvula.

being horked out the nostrils. Nasal regurgitation is more common with children, because they are often laughing while eating and because their swallowing mechanism isn't fully developed.

"Immature swallowing coordination" is the reason 90 percent of food-related choking deaths befall children under the age of five. Also contributing: immature dentition. Kids grow incisors before they have molars; for a brief span of time they can bite off pieces of food but cannot chew them. Round foods are particularly treacherous because they match the shape of the trachea. If, say, a grape goes down the wrong way, it blocks the tube so completely that no breath can be drawn around it. A child is better off inhaling a plastic barnyard animal or toy soldier, because air can be inhaled through its legs or around its rifle. Hotdogs, grapes, and round candies take the top three slots in a list of killer foods published in the July 2008 issue of the *International Journal of Pediatric Otorhinolaryngology*, itself a calamitous mouthful. Jennifer Long, a professor of head and neck surgery at the University of California, Los Angeles, went so far as to declare hotdogs a public health issue. A candy called Lychee Mini Fruity Gels has killed enough times for the U.S. Food and Drug Administration to have banned its import.

Every now and then a food comes along that is so difficult to orally process that even healthy adults without dysphagia have trouble getting it down. Sticky rice mochi, a traditional Japanese New Years food, kills about a dozen people every year—along with puffer fish and flaming cheese, the world's riskiest menu items.

The safest foods, of course, are those that arrive on the plate premoistened and machine-masticated, leaving little for your own built-in processor to do. They are also, generally speaking, the least popular. Mushy food is a form of sensory deprivation. In the same way that a dark, silent room will eventually drive you to hallucinate, the mind rebels against bland, single-texture foods, edibles that do not engage the oral device. Mush is for babies. Those who can, want to chew. The story of U.S. military rations bears this out. During World War II, when combat rations were tinned, meat hashes were a common entrée because they worked well with the filling machines. "But the men wanted something they could chew, something into which they could 'sink their teeth,'" wrote food scientist Samuel Lepkovsky in a 1964 paper making the case against a liquid diet for the Gemini astronauts. He summed up the soldiers' take on potted meat: "We could undoubtedly survive on these rations a lot longer than we'd care to live." (NASA went ahead and tested an all-milkshake meal plan on groups of college students living in a simulated space capsule at Wright-Patterson Air Force Base in 1964. A significant portion of it ended up beneath the floorboards.)

The only thing sadder than swallowing mush is not swallowing at all. Tube-feeding is a deeply depressing state of affairs. Rather than chew and spit out his food, Tom Little—the Irishman with the strictured esophagus—could have mashed it and pushed it directly into his stomach. In fact, he tried this, but without chewing, it "failed to satisfy." (Beer, however, was poured directly into the funnel.) Here's how badly people want to chew. Recall that dysphagia may knock out the reflex that repositions the larynx (voice box) to allow food into the esophagus. Jennifer Long told me these patients have on occasion asked to have their voice box surgically removed so they can swallow again. In other words, they would rather be mute than tube-fed.

Crispy foods carry a uniquely powerful appeal. I asked Chen what might lie behind this seemingly universal drive to crunch things in our mouths. "I believe human being has a destructive nature in its genes," he answered. "Human has a strange way of stress-release by punching, kicking, smashing, or other forms of destructive actions. Eating could be one of them. The action of teeth crushing food is a destructive process, and we receive pleasure from that, or become de-stressed."

I run this by René de Wijk when I get back to his house in the evening. He is slouched on the sofa, his frizzy hair falling in clumps on his forehead. His son sits between us, playing Assassin's Creed on the TV screen. A man in a cowled robe is doing some de-stressing, bludgeoning people and slicing them in two with a broadsword.

René agrees with Chen's assessment. "With crispy, it's so obvious that you're destroying the food in order to get your sensation. What is more marvelous than to control a nice structure with your mouth?" René doesn't know offhand of any studies on the psychology of crunchy food, but he promised to e-mail a colleague, Ton van Vliet, a food physicist who has devoted the past eight years of his career to a deeper understanding of crispy-crunchy.

The assassin bisects another citizen while René and his wife discuss the thermostat. The heating people have been out to fix it, and now they're coming back because it's on the fritz again. I point the toe of my boot at the TV. "That guy seems effective. Get him on board."

René looks at the screen. "He has his creed, he would *kill* the heating people!"

I was originally to have spent the afternoon with René in the Wageningen University Oral Lab. He had promised to wire me up to the articulograph and generate a 3-D profile of my chewing style, but he couldn't recall which sensor went where. I sat with a beard of colored wires hanging from my cheeks while René flipped through the manual. And then he had to leave for a meeting.

Nonetheless, he's been very effective at persuading other harried researchers to let me eat up their time. Ton van Vliet has agreed to meet us the following day at my home-away-from-home, the Restaurant of the Future.

V AN VLIET IS there when René and I arrive, sitting with his back to us at a table in the middle of the room. René recognizes the white hair. The longer strands appear to originate from a source at the back of the head and travel forward from there. All I can guess is that he walked here with a fierce wind at his back.

Van Vliet looks up as if from deep thought, a little startled, and extends his hand. He has a fine-boned face accented by an Amish-style beard and delicate-looking wire-rim glasses. I don't want to use the word *elfin,* in case it seems belittling, but it did come to mind.

Van Vliet wants to start me out with the basics of crispy-crunchy. We begin with nature's version, a fresh apple or carrot. "It's all bubbles and beams," he says, sketching networks of water-filled cells and cell walls on a sheet of my notepad paper. When you bite into an apple, the flesh deforms, and at a certain moment the cell walls burst. And there is your crunch. (Ditto crispy snack foods, but here the bubbles are filled with air.) "This is why fresh fruit is crisp, and also why it is a little bit juicy," says van Vliet. His voice is reedy and high-pitched, with a musical cadence.

As a piece of produce begins to decay, the cell walls break down and water leaks out. Now nothing bursts. Your fruit is no longer crisp. It is mealy or limp or mushy. The same thing happens with a snack food degraded by moisture: cell walls dissolve, air leaks out.

The staler the chip, the quieter. For a food to make an audible noise when it breaks, there must be what's called a brittle fracture—

a sudden, high-speed crack. "Like this." Van Vliet is drawing graphs again. As you bite down on a chip, energy builds and is stored. In a millisecond, the chip gives way and the stored energy is released, all at once. *Crack* is a superb onomatopoeia; the word sounds like the noise, and the noise *is* the fracture. (Crumbly foods, by contrast, break apart quietly because the energy isn't released all at once.)

Van Vliet reaches for a bag of puffed cassava chips René bought for us to use as props. He snaps one in two. "To get this noise, you need crack speeds of 300 meters per second." The speed of sound. The crunch of a chip is a tiny sonic boom inside your mouth. Van Vliet rubs his palms together to brush off the crumbs. This too makes a sound, dry like papers being shuffled. The Dutch winter is a brutal desiccant, to borrow from the language of snack foods.

René and I have been working our way through the props. He tilts the bag toward van Vliet, who waves it off. "I don't like chips and things."

René and I exchange a glance: Get *out!*

"I like *beschuit* . . ." He turns to me. "It's a Dutch toast that is round. We serve it when babies are born."

René wears an expression that FaceReader will have no trouble decoding. "Are you kidding me? It is *so* dry. I mean, you *cannot* move your tongue anymore! Really, I am hoping no more babies are born."

"It's very nice," insists van Vliet. "You have to put butter on it, and then honey on it."

I get up to look for some, but the restaurant has none.

Van Vliet juts his jaw. "Then that is not a good restaurant."

René leans in close to van Vliet, laughing. "It's a *very* good restaurant that takes care of its customers."

Moving along, van Vliet provides the answer I was looking for. Crispness and crunch appeal to us because they signal freshness. Old, rotting, mushy produce can make you ill. At the very least, it has lost much of its nutritional vim. So it makes sense that humans evolved a preference for crisp and crunchy foods.

To a certain extent we eat with our ears. The sound made by biting off a piece of carrot—more so than its taste or smell—communicates freshness. René told me about an experiment in which subjects ate potato chips while a researcher digitally altered the sounds of their chewing. If they muted the crunch or masked the higher frequencies, people no longer sensed the crispness. "They rated the chips as old even though the texture had not changed."

Van Vliet is nodding. "People eat physics. You eat physical properties with a little bit of taste and aroma. And if the physics is not good, then you don't eat it."

Crispness and crunch are the body's shorthand for "healthy." The snack-food empires have cashed in on this fact, producing crisp, crunchable foods that appeal to us but fail to deliver in terms of health and survival.

A good amount of thought appears to have gone into designing optimal crunch. "People like it most when it is around 90 to 100 decibels," says van Vliet. To achieve that, you need about a hundred bubbles bursting in rapid succession. "An avalanche of cracks in your mouth! To the ear it sounds like one sound, but in fact it is made up of more than one hundred sound bursts." This is achieved by messing around with the bubbles and beams—their size, their brittleness.

It's a marvel: such sophisticated physics in the service of junk food. I ask van Vliet which crispy-crunchy snack foods he has helped design. He wears a look that conveys both amusement and something dimmer. "Oh, the food companies are not using this science. They just make a product, give it to somebody, and say, 'How do you like it?'"

René confirms this. "They are so low-tech. They have *no* clue." It takes five to ten years for the discoveries of food physics to find their way into industry.

What is the point, then? For van Vliet anyway, the point is physics. Earlier, when I'd complained that the food-texture journals were "just a lot of physics," van Vliet seemed taken aback. "But physics is so nice!" It was as though I'd insulted a friend of his.

René cranes his neck toward the steam tables. "Can you stay for lunch, Ton?" It's 12:30 and all we've had are cassava chips. With his tongue, René works some free from a molar.*

Van Vliet considers this. "Well, I would have to tell my wife. You see I'm a good Dutch man, I go home for lunch every day! On my bicycle." In his eight years at Wageningen University, he adds, he has never tried the food in Restaurant of the Future. We are unable to tell if this is a yes or a no. René asks him whether he has a cell phone, to call his wife.

"Yes, we have one at home."

We let it drop. Later, walking to the parking lot, we glimpse van Vliet on a campus bike path, pedaling into the slanting snow.

---

* Technical term: toothpack.

# 8

# *Big Gulp*

## HOW TO SURVIVE BEING SWALLOWED ALIVE

*I*N THE COLOR plate that illustrates the Jonah story in my mother's Bible, the fisherman is halfway in the mouth of an indeterminate species of baleen whale. He wears a sleeveless red robe, and his hair, just starting to recede around the temples, is slicked back with seawater. One arm is outstretched in an effort to swim free. Baleen whales are strainer feeders. They close their mouth on a large gulp of ocean and use their tongue to push it forward through the vast comb of baleen, expelling the seawater and retaining small fish, krill, anything solid. It is a gentle, perhaps even survivable, way to be eaten. The prey is rarely much larger than a man's foot, however, and the whales are built accordingly.

"Baleen whales have very small gullets," says Phillip Clapham, a whale biologist with the National Oceanic and Atmospheric Administration. "They could not possibly swallow a hapless victim of God's wrath." But a sperm whale could. Its gullet is wide enough, and though it has teeth, it doesn't, as a matter of course, chew its food. Sperm whales feed by suction. Evidently quite powerful suction: in 1955, a 405-pound giant squid—six foot six minus the tentacles—was recovered intact from the stomach of a sperm whale caught off the Azores.

And then there is James Bartley. On November 22, 1896, the *New York Times* picked up the story of a sailor on the whaling ship *Star of the East* who disappeared in the waters off the Falkland Islands after a harpooned sperm whale, "apparently in its death agonies," capsized his whaleboat. Assuming Bartley had drowned, the rest of the crew set to work flensing the whale, which had by then finished up its agonies. "The workmen were startled . . . to discover something doubled up in [the stomach] that gave spasmodic signs of life. The vast pouch was hoisted to the deck and cut open and inside was found the missing sailor, . . . unconscious" but alive—after thirty-six hours inside the whale.*

Bible literalists seized upon the Bartley story. For decades, it turned up in religious tracts and fundamentalist sermons. In 1990, professor and historian Edward B. Davis, then at Messiah College in Grantham, Pennsylvania, did some fact-checking. His paper runs to nineteen pages and encompasses research that took him from the newspaper archives of the British Library to the history room of the Great Yarmouth public library. Short version:

---

* 1896 was a banner year for human-swallowing, or yellow journalism. Two weeks after the Bartley story broke, the *Times* ran a follow-up item about a sailor buried at sea. An axe and a grindstone, among other things, were placed in the body bag to sink the parcel. The man's son, frantic with grief, plunged overboard. The next day, the crew hauled aboard a huge shark with an odd sound issuing from within. Inside the stomach, they found both the father and the son alive, one turning the grindstone while the other sharpened the axe, "preparatory to cutting their way out." The father, the story explains, "had only been in a trance." As had, apparently, the *Times* editorial staff.

The *Star of the East* was not a whaler, and there was no whaling going on in the Falklands at that time. No one named James Bartley had been on the ship, and the captain's wife was certain no crew had ever been lost overboard.

Placing history aside, let's look at the digestive realities of the Bartley situation. If survival in the stomach were a simple matter of the size of the accommodations, any one of us could manage just fine. The forestomach of a killer whale, a far smaller creature, has been measured, unstretched, at five feet by seven feet—about as big as a room in a Tokyo capsule hotel, with a similar dearth of amenities. Figure 154 of *Whales*, by esteemed whale biologist E. J. Slijper, is a scale drawing of a twenty-four-foot killer whale and the fourteen seals and thirteen porpoises recovered from its stomach. The prey are drawn in a vertical lineup beneath the whale's belly, like whimsically shaped bombs dropping from a plane.

While a seaman might survive the suction and swallow, his arrival in a sperm whale's stomach would seem to present a new set of problems.* "Bartley's skin, where it was exposed to the action of the gastric juices, underwent a striking change. His face and hands were bleached to a deadly whiteness and the skin was wrinkled, giving the man the appearance of having been parboiled." Hideous. And, it turns out, bogus. The whale's forestomach secretes no digestive fluids. Hydrochloric acid and digestive enzymes are secreted only in the second, or main, stomach, and the passage between first and second is too small to admit a human.

While the absence of acid in the sperm whale forestomach shoots another hole in the Bartley tale, it lends some credence to

---

* I challenge you to find a more innocuous sentence containing the words *sperm, suction, swallow,* and any homophone of *seaman*. And then call me up on the homophone and read it to me.

the Jonah parable. Let's say the whale swallowed some air as it surfaced in pursuit of Jonah. Or let's fast-forward a few centuries and give him a scuba tank of air. Might the whale stomach under these circumstances be a survivable environment?

It might, if not for this: "Whales 'chew' their food with their stomachs," writes Slijper. Since sperm whales swallow prey whole, they need some other way to reduce it to smaller, more easily digestible pieces. The muscular wall of the forestomach measures up to three inches thick in some species. Slijper compares the cetacean forestomach to the gizzard in birds—an anatomical meat grinder that stands in for molars.

Would a man in a whale forestomach be crushed or merely tumbled? Is the force lethal or just uncomfortable? No one to my knowledge has measured the contraction strength of the sperm whale forestomach, but someone has measured gizzard squeeze. The work was done in the 1600s, to settle an argument between a pair of Italian experimenters, Giovanni Borelli and Antonio Vallisneri, over the main mechanism of digestion. Borelli claimed it was purely mechanical: that birds' gizzards exerted a thousand pounds of force, and with that kind of grinding going on there was no call for chemical dissolution. "Vallisneri, on the contrary," wrote author Stephen Paget in a 1906 chronicle of early animal experimentation, "having had occasion to open the stomach of an ostrich, had found there a fluid* which seemed to act on bodies immersed in it."

In 1752, a French naturalist devised a way to resolve the debate—and, unintentionally, address the inane whale-stomach-

---

* Vallisneri named the fluid *aqua fortis*—not to be confused with *aquavit*, a Scandinavian liquor with, sayeth the Internet, "a long and illustrious history as the first choice for . . . special occasions," like holidays or the opening of an ostrich stomach.

survival query of an American author two and a half centuries into the future. René Réaumur owned—or anyway, had access to—a small raptor called a kite. Like most carnivorous birds, the kite regurgitates a pellet of fur and feathers once it has finished with the digestible portions of its prey. This gave Réaumur an idea. He could hide in the kite's food a small tube carrying meat. The tube would keep the meat from being crushed by the gizzard, and mesh grates at either end would allow stomach solvents, if they existed, to enter and digest it. The kite's gizzard, taking the tube to be an unusually large, hard bone, would conveniently return it to daylight. If the meat in the regurgitated tube was dissolved, it meant some sort of fluid had done the work of digestion. Réaumur would eventually try this with a variety of barnyard birds. For our purposes, we are more interested in the fate of the tubes than that of the food. Those made from glass were smashed by the contractions of the gizzards, as were the tin tubes that replaced them. Réaumur had to use lead tubes that would withstand close to 500 pounds of pressure before they emerged from a gizzard uncrushed.

To get a sense of what that would feel like—what it would be like inside a gizzard or, by extension, a sperm whale stomach—I did a Google search on "500 pounds of pressure." That is, among other things, the maximum pressure exerted by the beak of a Moluccan Cockatoo, a bird that can bite off a man's finger. It's the force exerted by the footfall of a 130-pound person, which means that being inside a gizzard feels like me stepping on you, perhaps in my haste to escape your cockatoo. And, finally, the American Automobile Association tells us that 500 pounds is the force with which an unrestrained ten-pound dog will hit the windshield in a fifty-mile-per-hour head-on.

And a sperm whale's forestomach muscles are presumably more powerful than those of a turkey gizzard. I'd say your chances of surviving in a sperm whale stomach are slender. I'd say you're better off with the Chihuahua in the crashing pickup truck.

The biblical account of Jonah's travails does not actually use the word *whale*. It says "big fish." University of California, Santa Cruz, biologist Terrie Williams once had occasion, as they say, to open the stomach of a sixteen-foot tiger shark. It happened while she was working in Hawaii. A woman had been killed while swimming, not far from where the shark had been caught, and Williams was called in to see if pieces of her might be found inside it. Instead, Williams found three full-grown, manhole cover–sized, intact green turtles, all facing forward. "They never saw it coming. All they knew was like, 'I'm swimming around and it's blue, it's Hawaii, how great is this . . .' And the next thing they see is this huge mouth shutting." And shark stomachs, unlike sperm whale forestomachs, secrete gastric acids and enzymes. Williams thought that the turtles, withdrawn into their protective shells and able to store oxygen in their muscles, might have survived a half day or so.

What about a scuba diver in a wetsuit with a tank of oxygen? How long could he survive in a tiger shark? Christiananswers.net puts forth an intriguing digestive loophole that, were it true, would have worked in his—or for that matter, Jonah's—favor: "As long as the animal . . . swallowed is still alive, digestive activity will not begin."

THIS PERSISTENT BIT of digestive bunk can be traced to eighteenth-century Scottish anatomist John Hunter, an otherwise estimable scientist who more or less invented modern surgery. In the course of hundreds of dissections, Hunter would come across cadavers with mysterious lesions in the stomach wall. He first assumed, reasonably enough, that the lesions had been the cause of death. But the condition was turning up even in vigorous young men killed in brawls, including one man done in by a blow to the head with an iron poker. In this case, too, the man's stomach was dissolved clear through, Hunter noting that the contents of his supper—cheese, bread, cold meat, and ale—had spilled into the body cavity. There are several things one might take away from this case: that pub fare has changed little in two hundred years; that the owners of drinking establishments would do well to keep the fireplace tools behind the bar. Hunter came away with the realization that the inexplicable lesions he'd been seeing were not disease but auto-digestion. The stomach tissue, he noted, was damaged in the same way the digested cold cuts were. In other words, the stomach, at death, begins to digest itself.

This raised the question, What keeps it from doing so while the person is alive? Hunter's explanation—and the source of the Christiananswers.net piffle—was that living tissue exudes some sort of vital force field that protects it. "Animals . . . possessed of the living principle, when taken into the stomach, are not the least affected by the powers of that viscus . . . ," stated Hunter in a 1772 text. Ditto humans taken in: "If one conceive a man to put his hand into the stomach of a lion, and hold it there," wrote Hunter in a separate text, ". . . the hand would not in the least be digested." A small and temporary consolation, it must be said.

French physiologist Claude Bernard didn't buy it. Bernard took some animals into the stomach. The year was 1855. The stomach belonged to a live dog and had been given a fistulous opening similar to the one that had enabled William Beaumont to spy on the digestive activities of Alexis St. Martin a few decades (and chapters) earlier. Bernard restrained the dog and then "introduced," through the fistula, the hindquarters of a frog. After forty-five minutes, the frog's legs were "largely digested"—nothing new to a Frenchman, except that here the frog was still alive. The experiment, concluded Bernard, "shows that life is not an obstacle to the actions of gastric juices." And that cruelty was not an obstacle to the actions of Claude Bernard.*

In 1863, English physiologist Frederick W. Pavy extended Bernard's findings to mammals. In keeping with the French market-day theme, Pavy selected a rabbit. He inserted one of its ears into the stomach of yet another fistulated dog while it was digesting a meal. Four hours later, a half inch of the tip was "almost completely removed, the small fragment only being left attached by a narrow shred to the remainder of the ear." Again, digestion had proceeded unthwarted by the "living principle" or any sense of decency.

So Hunter was wrong. No vital force exists to protect a living being from the effects of the gastric secretions. Why is it, then, that stomachs don't digest themselves? Why do one's stomach juices handily digest haggis or tripe but not the very stomach that secretes them?

It's something of a trick answer. In fact, stomachs can digest themselves. Gastric acid and pepsin digest the cells of the stomach's protective layer, or mucosa, quite effectively. What no one in Hunter's day realized is that the organ swiftly rebuilds what it breaks down. A healthy adult has a new stomach lining every three days. (More clever stomach tricks: key components of gastric acid are secreted separately, lest they ravage the cells that man-

---

* At some point during the experiment, or possibly the follow-up, wherein a live eel was pushed into the stomach and left with "just its head outside," or one of the dozens of other vivisections, Bernard's wife walked in. Marie Françoise "Fanny" Bernard—whose dowry had funded the experiments—was aghast. In 1870 she left him and inflicted her own brand of cruelty. She founded an anti-vivisection society. Go, Fanny.

ufacture them.) The stomachs of John Hunter's cadavers managed to burn holes in themselves because the mucosa-producing machinery shuts down at death. If someone dies in the midst of a meal—particularly in a warm clime, where weather stands in for body heat—the digestive juices continue to act though the restoration work has stopped.

If you must spend time in a digestive organ, I recommend the penguin stomach. Penguins can shut down digestion by lowering the temperature inside their stomach to the point where the gastric juices are no longer active. The stomach becomes a kind of cooler to carry home the fish they've caught for their young. Penguins' hunting grounds may be several days' journey from the nest. Without this handy refrigerated mode, the swallowed fish would be completely digested by the time the adults get back—"like going shopping and eating everything you bought on the way home," as marine biologist Terrie Williams put it.

ONE REASON THE notion of living principle gained traction in John Hunter's mind was that it offered a medical explanation for stomach snakes. As far back as Babylon and ancient Egypt, people had been coming to doctors with complaints of reptiles or amphibians living inside them. The malady hit an especially brisk stride in the late eighteenth century. "Thence it is," wrote Hunter in his 1772 paper on the living principle, "that we find animals of various kinds living in the stomach, or even hatched and bred there." Through the end of the century and likely beyond, biologists of imposing stature—not just Hunter, but also Carl Linnaeus—believed that frogs and snakes could live in humans as parasites, nourished by daily deliveries of swallowed food. Medical historian and author Jan Bondeson tracked down some five dozen case reports in medical journals from the seventeenth, eighteenth, and nineteenth centuries. Eighteen involved lizards or salamanders; seventeen posited snakes; fifteen claimed frogs; and twelve toads.

Despite the varied taxonomy and geography represented by these cases, the basic premise is more or less the same. The patient, vexed by odd sensations or pains in the abdomen, suddenly remembers a visit to the country. While walking home at night, the tale typically goes, he stops to drink from a pond—or marsh or rivulet or spring. It's nighttime, and he cannot see what he is swallowing. Or he is drunk and does not notice. Sometimes he believes he swallowed eggs, other times the actual animal. In a few instances, the person lies down to sleep or passes out, whereupon some elongate, cold-blooded creature slithers down his esophagus and into his gut.

What cements the delusion, in the patient's mind, is the timely sighting of an animal in the chamber pot. "When at stool, she had unusual pain in the rectum, and afterwards she thought she perceived something moving in the pot," reads a typical case report, from 1813. Often the patient has been given a cathartic to relieve his symptoms. As here, in the 1865 case report of a stomach slug: "The patient had taken an injection per anum* . . . and immedi-

---

* Meaning "by way of the anus." "Per annum," with two *n*'s, means "yearly." The correct answer to the question, "What is the birth rate per anum?" is zero (one hopes). The Internet provides many fine examples of the perils of confusing the two. The investment firm that offers "10% interest per anum" is likely to have about as many takers as the Nigerian screenwriter who describes himself as "capable of writing 6 movies per anum" or the Sri Lankan importer whose classified ad declares, "3600 metric tonnes of garlic wanted per anum." The individual who poses the question "How many people die horse riding per anum?" on the Ask Jeeves website has set himself up for crude, derisive blowback in the Comments block.

ately afterward something attracted his attention by moving about under his clothes."

The likelier chain of events, of course, is that the creature had been in the pot or the bed, unnoticed, all along. And that the authors who wrote these papers were either lazy thinkers or, equally possible, crafty career opportunists. Cases like these, taken at face value, were irresistible medical curiosities; reports of them were sure to be published in medical journals and newspapers of the day, spreading the physicians' names and boosting their status.

Then again, to be fair, some of the details conspired to lend credence to the claims. Like the contemporary urban myth, tales of stomach frogs and "bosom serpents" persisted because they have truthiness. Few would believe a story about a man with a mammal alive in his digestive tract—though Bondeson tracked down one instance of a stomach mouse—but an indwelling frog has biological plausibility. Sideshow regurgitators used frogs because they can absorb oxygen from water through their skin. Swallow a frog in a large glass of water, and it will survive—at least through the end of the act.

Cold-blooded animals in general have lower metabolic needs. Because they're not using food energy to heat themselves, they manage with less. Some frogs all but shut down in winter. "I wouldn't be surprised if live frogs were gutted out of bass in winter, by fisherman," wildlife biologist Tom Pitchford told me. But a human belly isn't cold. It's tropical. Around 1850, in Germany, physiologist-zoologist Arnold Adolph Berthold, seeking to put an end to stomach-frog folly, put some northern European species of frogs and lizards in body-temperature water. The adults died, and the spawn putrefied.

That snakes top the list is not surprising. On top of their over-all cold-blooded hardiness, they seem to have a special knack for enduring gastrointestinal confinement. Phillip Clapham, the whale biologist I pestered at the start of this chapter, related the story of Gracie, a Doberman mix who once vomited a two-foot garter snake onto Clapham's dining room floor during a dinner party. As he tells it, his wife at the time, assuming the snake was dead, picked it up in a wad of paper towels and then "nearly dropped it when its little forked tongue came out." Clapham insists Gracie hadn't been outside for at least two hours. "It had been in there quite a while."

University of Alabama snake digestion researcher Stephen Secor once watched a king snake regain consciousness after somewhere between ten and twenty-five minutes inside another king snake. He had put the two in the same tank, not realizing one species considered the other dinner. Secor left the room, and when he returned, dinner was "down in." He pulled them apart, and was relieved and surprised to note that dinner still had a heartbeat.

Nonetheless, a brief sojourn is different from permanent immigration. More reputable doctors of yore recognized stomach snakes for what they were: delusions inspired by gastric symptoms. The underlying condition was typically mundane: ulcer, lactose intolerance, intemperance, gas. You could often tell what was going on from the patients' descriptions of their tenant's habits. Andrew S.'s snake acted up whenever he drank alcohol or milk. "He will never allow me to drink whiskey," S.'s physician Alfred Stengel quotes him as saying in the 1903 paper "Sensations Interpreted as Live Animals in the Stomach." "He hates that worse than anything else." The stomach snake of a woman in Castleton, Vermont, circa 1843, was most active after "any considerable indulgence in gross food."

Occasionally there was nothing wrong at all, just the ordinary grumbling and gurgling—the borborygmi—of the gut. The surgeon Frederick Treves, writing in the late 1800s, described five cases of patients complaining of wriggling movements or of live snakes inside them. Upon operating and finding nothing beyond the normal motions of a healthy digestive tract, he coined a term: "intestinal neurosis." It exists today, minus the snakes. One gastroenterologist told me about a sad soul who wandered the motility clinics of North America with a video of himself in his underwear, pennies stacked on his abdomen to show the alarming motions of his (perfectly normal) intestines.

Sometimes a patient would manage to capture the alleged tormentor and bring it in to show the doctor. While some physicians kept the animals for display in cabinets of curiosity—or, on occasion, as pets—those of a more scientific bent recognized an opportunity for forensic fact-checking. Jan Bondeson relates a famous case from the seventeenth century of a twelve-year-old who complained of abdominal cramping and, over an unspecified span of time, allegedly vomited twenty-one newts, four frogs, and "some toads." One of the youth's physicians had the bright idea of dissecting the amphibians' stomachs. If the story were true, the food inside the little stomachs should reflect the creatures' gastrointestinal habitat. Instead the stomachs contained half-digested insects. In 1850, Arnold Adolph Berthold, our man of the putrefied frog spawn, approached curators at German medical museums whose collections included reptiles and amphibians allegedly vomited or excreted after years of residence in a human digestive tract. Here again, when specimens' stomachs were opened, many were found to hold insects in various stages of dissolution.

The most directed experimental debunking was carried out by J. C. Dalton, a physiology professor at the College of Physicians and Surgeons of New York. Over a span of several months in 1865, Dalton had twice been visited by flummoxed colleagues bearing "discharged" slugs in jars of alcohol. One was said to have come from a boy who had been suffering three weeks from diarrhea. The usual narrative ensued: "It was during this diarrhea that the slugs were passed. On that day, the mother, on removing the clothes from the child after a fecal evacuation, found among them one of the animals, alive and moving." She assumed he'd inadvertently consumed slug eggs while eating garden greens on a family visit in the countryside, where the boy had "passed a part of the summer"—an alarming verb choice under the circumstances.

Dalton was dubious. "I accordingly thought it worthwhile to institute some experiments, with a view of ascertaining how far such a thing might be possible." Garden slugs were procured from a neighborhood lettuce bed. An assistant held a dog's mouth open while Dalton placed four slugs, one at a time, at the far back of its mouth to get it to swallow without chewing. An hour later, Dalton took out his scalpel. He found "no recognizable traces of slug" anywhere along the dog's alimentary canal. In subsequent experiments, just fifteen minutes rendered a slug "somewhat softened" and a salamander "exceedingly soft and flaccid," and both dead.

"It is a curious psychological phenomenon," wrote Dalton, "to witness the thorough confidence . . . and the fullness of detail with which intelligent persons will sometimes relate these stories. . . . When the accounts come to us second hand, we can always make abundant allowance for the natural growth of wonders, in passing from mouth to mouth. But even when the facts stated are those which came under the relator's own observation, the discrepancy between his convictions and the truth may sometimes be equally remarkable."

Wise words, no less applicable today. It is 2011 as I write this, and the story endures. Only now the lizards and frogs are on the outside.

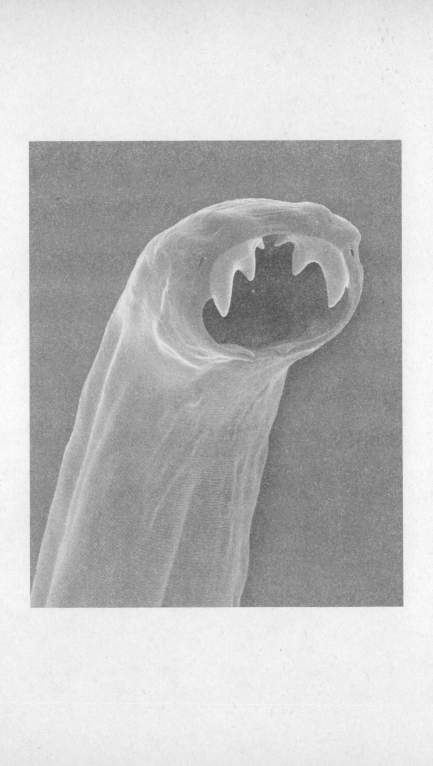

# 9

# Dinner's Revenge

## CAN THE EATEN EAT BACK?

THE DARKLING BEETLE, small and shy with an understated matte-black carapace, is better known as its adolescent self, the mealworm. Mealworms and their darkling cousins the superworms are popular "live feeders"—food for pet reptiles and amphibians that won't eat prey that's already dead. For years, a disconcerting rumor has bounced around the "herp" (as in, herpetofauna) community. Heed the words of Fishguy2727, posting on Aquaticcommunity.com: "I have talked to a number of people who have FIRST-HAND watched with their own eyes as the animal ate a mealworm, . . . and within ten to twenty seconds the mealworm is chewing out of the animal's stomach."

I heard about the phenomenon SECOND-HAND from wildlife biologist Tom Pitchford. The mealworm came to mind when I asked Tom whether he knew of any nonparasitic creature that could survive in a stomach for any length of time. He had heard that some online herp forums recommend crushing mealworms' heads prior to serving. "While the insect is in its death throes, the lizard will come over and eat it."

Mealworm ranchers scoff. "This is an old wives tale," says Wormman.com. The owner of Bassetts Cricket (and mealworm) Ranch told me that a slice of carrot, for a mealworm, is a two-day project. "They can't eat out," he said. (Though obviously enough people worry about it that it has its own verb form.) But mealworm sellers have a financial stake in the matter. What do reptile and amphibian dealers say? Carlos Haslam, manager of the East Bay Vivarium, a reptile and amphibian store not far from my home, told me that in his forty years in the business, he has not seen the phenomenon nor heard a customer report it happening. He pointed out that lizards chew their food before swallowing. Frogs don't, but lizards do. And most of the stories are about lizards. Fishguy2727 takes no comfort. "Just because 1,000 people have not had it happen to them does not mean it is impossible. There is no doubt that this can happen."

As so often is the case with apocryphal tales like this, finding someone who *knows someone* who's seen it is easy. Less easy is tracking down an actual eyewitness. One who claims to have seen is John Gray, the animal care technician at the Tracy Laboratory at the University of Nevada, Reno. His boss, Richard Tracy, is a physiological ecologist. He predicts hotspots of future extinction, with reptiles and amphibians as his focus. Eighteen lizards, forty toads, and fifty frogs are under John Gray's care, but he has not seen it happen to any of them. It happened to a fence lizard he caught in his backyard as a twelve-year-old. He recalls feeding a superworm to his new pet in the evening, and finding the lizard dead the next morning with the superworm "hanging out of its side."

Tracy is skeptical. He has a theory that the story took root in the public's consciousness with the 1979 release of *Alien*, a film in which the title character hatches inside one of the crew and breaks through the skin of the man's abdomen during a meeting. He questions Gray's memory. Who can recall, with dependable accuracy, the details of an event that happened thirty years ago? One of the mealworm's natural behaviors is to crawl underneath things. "Mealworms prefer darkness and to have their body in contact with an object," says the University of Arizona Darkling Beetle/Mealworm Information sheet, under the heading "Interesting Behaviors." The sheet's authors make no mention of mealworms eating their way out of stomachs, which would, you'd think, qualify as interesting behavior. As with the post-laxative stomach slug and snake sightings of yesteryear, it seems more likely that the worm was already on the scene, seeking darkness and framed by happenstance.

However, like most people who work with captive reptiles and amphibians, Tracy has trouble completely dismissing the stories. He's going to do what experimental biologists do in situations like this: experiment.

PROFESSOR TRACY HAS borrowed an endoscope. It is slimmer than most because it was designed to look up urethras. The scope belonged to a urologist whose daughter studied tortoises at the University of Nevada. He lent it to her to look inside tortoise burrows, and she has lent it to Tracy to watch mealworms inside stomachs. What goes around comes around, and up and in and through.

Tracy has no funding for the experiment, just enthusiasm. He calls up colleagues and acquaintances and tells them what he's fixing to do, and they jump on board with offers to help. Walt Mandeville, the university veterinarian, has volunteered to do the sedating. Tracy's grad student Lee Lemenager will be manning the endoscope. Lee has the kind of face that children draw when they first begin to draw faces, everything round and benign. Earlier in the day when he dripped gastric acid on a superworm, it seemed like a friendly thing to do.

"And this is Frank and Terry, from OMED," says Tracy as two more men show up in the lab. OMED of Nevada sells used medical equipment. "They lent us tens of thousands of dollars of video equipment that is forty years old and probably worthless. Welcome!" Tracy is one of those supremely likable professors whom students keep in touch with long after graduation. The back wall of the Tracy Laboratory is covered with photographic portraits he has taken of his grad students. His white hair suggests he may be closing in on retirement, but it is difficult to imagine him golfing or watching daytime television.

Tracy holds a bullfrog in sitting position while Lee feeds the scope into its mouth and down to the stomach. We aim to spy on a superworm swallowed less than two minutes ago. The endoscope, which is a flexible tube of fiber optics with a tiny camera and light at the end, is hooked up to a closed-circuit video monitor so that everyone can watch, and Tracy can film, what's happening inside the stomach.

The frog is sedated but awake. It glows like a decorative table lamp, the kind that sets a mood but is not sufficient to read by. The screen on the monitor is solid pink: the view from inside a well-lit frog stomach. You don't expect any part of a frog to be pink, but there it is, pink as Pepto-Bismol.

And then suddenly: brown. "There he is!" Lee focuses down on telltale bands of brown, tan, and black. The superworm is not moving. To see whether it's even alive, Walt the veterinarian inserts a pair of biopsy forceps through the makeshift speculum that Lee slid down the frog's esophagus at the beginning of the experiment. The jaws of the forceps gently squeeze the superworm's midsection. It squirms, eliciting a spontaneous Broadway chorus: "It's alive!"

"Is it chewing?" someone asks. As if by director's cue, all heads lean in.

"That's the tail," says Walt the vet. Walt has a keen observational eye, honed by a span of years as a poultry inspector ("4.8 seconds per bird").

Lee pulls back on the endoscope and works it over to the other end. The superworm's mouthparts are still. Nothing is moving. Walt tells us about a phenomenon he calls the "blanket effect." To calm a wild horse prior to treating it, a vet may herd the animal into a narrow chute lined with packing peanuts that gently presses in on its sides. It is the same principle behind swaddling an infant or hugging a distraught friend or dressing a thunder-phobic dog in an elasticized Thundershirt, available in pink, navy, and heather gray. Mercifully, stomach walls seem to act as a mealworm Thundershirt.

Before the superworm was presented to the frog, Lee looped a thread around its middle and secured it with surgical glue, so he could retrieve it later. Now that time has come. The frog surrenders its lunch seemingly without concern, and the superworm is left in a petri dish to recover. John Gray goes to get a chuckwalla, placing the superworm back behind the lizard's teeth. Same result. The superworm quickly goes still but does not die.

One thing is clear from these experiments. Mealworms are not much troubled by gastric—that is, hydrochloric—acid. Many people, including myself when I began this book, think of hydrochloric acid more or less the way they think of sulfuric acid, the acid of batteries and drain cleaners and hateful men who wish to scar women's faces. Sulfur likes to bind with proteins, radically altering their structure. If that structure is your skin, you come away from the experience disastrously altered. Hydrochloric acid isn't as caustic.

For me the confusion can be traced to the movie *Anaconda*, the scene in which the giant snake rises from the water to regurgitate Jon Voight's character, his face melted like wax. Some time back, I visited the lab of my favorite snake digestion expert Stephen Secor, the technical consultant on *Anaconda*. I told him I wanted to experience gastric acid, to get a sense of what it might feel like to be alive inside a stomach. He made me promise not to tell his wife, who oversees safety protocol for the university's labs, and then he took a bottle of hydrochloric acid off a shelf and put a dab—five microliters—on my wrist. I braced for sharp heat, as from a drop of scalding water. It was a full minute before I felt anything at all, and then only a weak itch. He added another drop. At three minutes, the itch turned to mild irritation, which held more or less steady for twenty minutes, then faded to nothing. It left no mark.

But stomachs secrete more than a single drop of hydrochloric acid. And they keep on secreting, readjusting the pH as the digesting food buffers the acid. My guess is that the situation inside an actively secreting stomach lies somewhere between what occurred on my wrist and what happened to the Japanese factory worker who fell into a tank of hydrochloric acid seven feet deep. The case report states that his skin turned brown and the delicate tissue of his lungs and digestive organs underwent "dry coagulation necrosis." Burning—whether from acid or from heat—denatures proteins. It changes their structure. It is denaturing that solidifies the boiling egg, that curdles milk, that distorts the burn victim's skin.

Inside a stomach, hydrochloric acid denatures edible proteins, making them easier for digestive enzymes to break down.

The effects of gastric acid are insidious but far from instantaneous, especially if the eaten entity is, like a superworm, protected by an exoskeleton. Crabs vomited after three hours in the stomach of the Asian crab-eating snake *Fordonia leucobalia* have been known to stand up and run away. I have an eyewitness for this: University of Cincinnati biologist Bruce Jayne. Jayne had "gently massaged" the snakes' bellies to get them to surrender what they'd eaten, so he could tally it for his research. Because you can't just ask them.

But without Bruce Jayne to massage the belly, without Lee Lemenager to pull the surgical thread, without God making the whale regurgitate, there would seem to be no way out.

Parasites are the exception. "Parasites bore all over the place," says Professor Tracy. Some are equipped with a boring tooth, like a drill bit installed on the top of the head. "That's what they've evolved to do. But these are mealworms, for crying out loud." Larvae burrow, but they don't bore. "How the hell would they know to tunnel out?" Walt the vet agrees. He is off and running with a story about the giant kidney worm, a parasite that bores out the entire organ and then exits the body through the urethra. He jerks his elbow toward the endoscope. "You could watch it coming with that scope."

TRACY IS GOING to give the superworms one last chance, the best possible chance, to see if they can chew their way to freedom. They will be put inside a dead stomach—one with no secretions and no muscle contractions.

Where do you find a stomach on a Thursday afternoon in Reno?

"Chinatown?" suggests someone.

"Costco?"

"Butcher Boys." Tracy pulls his phone from a pocket. "Hello, I'm from the university"—the catchall preamble for unorthodox inquiries. "I'm wondering, is there any chance at all we could get a fish stomach from you?" Tracy waits while the man goes to ask someone and/or make twirling finger motions at his temple for the benefit of his coworkers. The lab falls quiet. The feeder crickets chirp in the next room. "No stomachs of anything? No. Okay."

John Gray lifts his head and says, in his quiet way, "I've got a dead leopard frog in the freezer."

Everyone takes a break while Gray goes to defrost his frog under a warm tap. Walt entertains us with talk of an alternative-medicine experiment going on at the medical school—healers practicing Reiki on mice. Tracy walks next door to get a toad to show me, a new species he discovered doing fieldwork in Argentina. He returns with it in a glass dish, cradled against his belly. He looks like a kid standing in the kitchen with his cereal bowl. It's a nice toad, less warty than some. I tell him this, and he seems pleased. "You could be the first person to like this species." Second, I'm pretty sure.

"You could be the last too," says Lee, more of a frog guy.

Gray rejoins the group with the defrosted leopard frog, now pinned in a dissecting tray. Lee snips up the midline of the belly and peels back the flaps of skin as if they were stage curtains. Professor Tracy slides a superworm into the stomach.

The 1925 essay "The Psychology of Animals Swallowed Alive" opens with the author sitting "in quiet contemplation digesting after dinner" and wondering whether animals that swallow their prey live* are "worried by the acrobatic effects of victims trying to escape." If this leopard frog were alive, if frogs have the neurological wherewithal to worry, then the answer must be yes, they sometimes worry. The mealworm, with obvious worries of its own, animates the frog stomach like a sock puppet, arcing and straightening and squirming in the snug pink sac for fifty-five seconds. Then it stops completely. "Blanket effect," says someone.

The superworm is extracted and set aside. Like the others, it is motionless but not dead. And as with all the earlier entrées, this one will wake up after half an hour or so outside the stomach and appear to be fully recovered. A second worm is left in place overnight, to rule out the possibility that superworms can shrug off the blanket effect and resume their efforts to escape. It is dead by morning. "There is no way in my mind that they can eat their way out of stomachs," states Tracy.

Walt is not as sure. He was impressed by the vigor of the

---

* Those of you who swallow oysters without chewing them may be curious as to the fate of your appetizers. Mollusk scientist Steve Geiger surmised that a cleanly shucked oyster could likely survive a matter of minutes inside the stomach. Oysters can "switch over to anaerobic" and get by without oxygen, but the temperature in a stomach is far too warm. I asked Geiger, who works for the Florida Fish and Wildlife Research Institute, about the oyster's emotional state during its final moments inside a person. He replied that the oyster, from his understanding, is "pretty low on the scale." While a scallop, by comparison, has eyes and a primitive neural network at its disposal, the adult oyster makes do with a few ganglia. And mercifully, it is likely to go into shock almost immediately because of the low pH of the stomach. Researchers who need to sedate crustaceans use seltzer water because of its low pH. Geiger imagined it would have a similar effect on bivalves. But you might like to chew them nonetheless, because they're tastier that way.

superworm's struggle. "What if there were a weak spot in the stomach?" Might it be possible to escape a stomach by rupturing it with an especially forceful squirm?

That appears to be what was depicted in a photograph that went viral in 2005, of a dead python in a Florida swamp with the tail and hind legs of an alligator sticking out of its side.

"That's what everyone was saying: that the alligator kicked its way out," Stephen Secor told me. Secor had been flown out to the scene by a National Geographic television production team, who had hired him as an on-camera expert for a one-hour special spawned by the chimerical remains. Secor knew before he arrived that the dinner-kicking-its-way-out scenario was extremely unlikely. Pythons kill their prey before eating it.* "And there's no way stuff can move once it's inside there."

There was in fact a weak spot. Secor pointed to a printout of the photograph I'd brought with me when I visited his lab in late 2010. Two-thirds of the way down the python's exterior is a patch of black (dead) tissue—a poorly healed wound from some earlier incident. The rupture of this wound, Secor thinks, was caused by an alligator, let's call him alligator B, who attacked the python while he was digesting alligator A. The python broke open at the poorly healed wound, and A popped out. So it wasn't, at the end

---

* *How* remains a matter of debate. I had heard that pythons suffocate prey by tightening on its exhale and preventing further inhales. Secor says no; prey passes out too quickly for that to be the explanation. "You'd still have oxygen circulating in the blood, like you're holding your breath." He thinks it's more likely that the constriction shuts off blood flow, more like strangulation than suffocation. An experiment was planned at UCLA but nixed by the animal care committee. Secor would volunteer himself. "I think we'd all like to have a giant snake constrict us in a controlled situation and see what happens— could we still inhale?" It's possible he's a little nuts. But in a good way.

of the day, a case of dinner exacting revenge from within. Just another dog-eat-dog day in the Everglades.

THE OTHER THEORY Stephen Secor debunked for the National Geographic program was that the alligator dinner was so enormous the python simply burst. "That," he said, pointing to the meal in the famous photograph, "is nothing." The python is built to accommodate prey many times wider and bulkier than itself. The esophagus is a thin, pink stretchable membrane, a biological bubble gum. Secor went over to his computer and pulled up a slide of a python engulfing the head, neck, and shoulders of an *adult kangaroo*. This was followed by a shot of a python with three-quarters of a gazelle "down in," with only the hips and rear legs remaining al fresco. Pythons use their muscular coils to pull the prey apart, like taffy, so it's narrower and easier to get down. And they don't swallow in a single peristaltic wave of muscle contraction, as we do. They do what's called a "ptergoid walk." They inch their jaws along on the prey like marines on their bellies, moving forward by the elbows, left, right, left.

The other reason Secor could dismiss the bursting-stomach theory is that he knows exactly how much pressure that would take. "We sealed off the cloaca of a dead python and inserted an air line down the esophagus." Probably much like you at this moment, Secor was "sick of listening to people talk about pythons bursting." I would give you the citation for his experiment, but Secor did not publish a paper. It was "just a fun thing." He pointed to my printout of the python-alligator photo. "It was a lot more pressure than could be generated from this."

Biologists have a term for stretchy, accommodating digestive equipment: compliant. *You're planning on taking down an ibex? Yes.*

*No problem. I can handle it.* The compliant stomach is a physiological larder, a storage unit for the food that will sustain an animal over the days or weeks when prey are scarce or it's off its game. It is the stomach of feast-or-famine. "The predator has a very compliant stomach," says David Metz, a gastroenterologist with the Hospital of the University of Pennsylvania who has studied people who compete in eating contests. "Think of the lion after the big meal, with its huge, distended belly. They can lie in the sun for the next few days, letting it all slowly get digested." When you occupy the top spot on the food chain, you are free to lounge around with little concern over someone larger and stronger jumping you and eating you. The lion falls prey only to humans, in the form of hunters—and the occasional Mesopotamian vivisectionist.

In a 2006 issue of the *Lebanese Medical Journal*, Farid Haddad details the efforts of Ahmad ibn Aby al'Ash'ath, a court physician in Iraq circa A.D. 950, to document the compliancy of a lion's stomach. In his opening paragraph, Dr. Haddad notes that *'ash 'ath* means "disheveled." It seems an unlikely name for a royal physician, but a brief spin through the man's writings sheds some light: "When food enters the stomach . . . , its layers get stretched; I observed this in a live lion which I dissected in the presence of Prince Ghadanfar. . . . I proceeded to pour water in the lion's mouth and continued to pour jug after jug in its throat; and we counted until the stomach filled up with about [5 gallons]. . . . I then cut open the stomach and let the water out; the stomach shrank and I could see the pylorus. God is my witness."

The agriculturally informed reader may be unimpressed by the five-gallon capacity of the lion's tank. A cow's rumen—the largest of its four stomach compartments—is the size of a thirty-gallon trash can. Why should this be, when all a ruminant needs to do to get dinner is lower its head and graze? When food carpets the land from hoof to horizon, famine isn't a concern. So why the massive intake? The answer lies in the relatively low nutritional value of the ruminant diet. It is not merely the size of the cow's rumen that resembles a garbage can, it is the contents. The first place I visited for this book was the University of California at Davis, where animal science professor Ed DePeters and his colleagues test organic waste by-products to see whether they might make good cattle feed. With the help of a fistulated cow, DePeters has tested the digestibility of almond hulls, pomegranate scrap, lemon pulp, tomato seeds, and cotton seed hulls. He is a modern-day William Beaumont, lowering mesh bags of experimental foods into the rumen, and then pulling them out by a string at intervals to see what remains. The day I visited, they had been testing prune pits from nearby Yuba City, "the prune capital of the world."*

Cows, by virtue of the plentiful and varied bacteria in their rumen, are able to derive energy from things that would pass through a human undigested. The prune pit has a hard, nutritionally blank hull, but the embryo inside provides protein and

---

* Excuse me, I mean the Dried Plum Capital of the World. The change was made official in 1988, as part of an effort to liberate the fruit from its reputation as a geriatric stool softener. Yuba City has Vancouver, Washington, to blame for that. The original Prune Capital of the World, Vancouver was the home of the Prunarians, a group of civic-minded prune boosters who, back in the 1920s, touted the laxative effects of dried plums. The Prunarians also sponsored an annual prune festival and parade. A 1919 photo reveals a distinct lack of festiveness and pruniness. Eight men in beige uniforms stand in a row across the width of a rain-soaked pavement. A ninth stands on his own just ahead of the row, similarly attired. Presumably he is their leader, though you expect a little foofaraw from an entity known as the Big Prune. Or the Big Dried Plum, as Yuba City would like you to call him.

fat. Rumen bacteria can break down the hull and free these nutrients, though it takes them a few days. DePeters showed me one of the mesh bags. "Sometimes I put a midterm exam in there," he said. Cows can't digest wood pulp. "I tell my students, 'The cow didn't digest that material any better than you did.'"

"We've done cloth from a plant in Petaluma that was making cotton towels. All the small linters that didn't get into the towels? You can feed 'em. They can break it down. They get energy from it. It's just slower." As with hay and grass, it takes a sizable serving of tea towel for a cow to get its RDA—hence the enormous volume of the rumen. DePeters speculates that there's another reason for the huge capacity of the rumen. Ruminants graze on the open plain, easily visible and vulnerable to predators. "So they'll go out and graze and take in a lot, then go and hide somewhere to ruminate and digest." The rumen is a built-in to-go box.

DePeters took me to visit one of the fistulated cows. Escorted by an entourage of large flies, we made our way through a grid of muddy corrals. I was in kitten heels and a skirt, a fact from which DePeters, in filth-encrusted rubber boots and worn T-shirt, derived lasting merriment. DePeters is tanned and tall, with a wiry build. His hair is the same reflective silver of the screeching aluminum gates. It works well with his eye color, the deep dusty blue of scrub-jay plumage.

Cow 101.5 was getting a hose bath from one of DePeters's students, Ariel. Ariel and her array of piercings posed a welcome challenge to the stereotype of the conservative male ag major. We stood by, watching and waving away flies. I like the look of cows: the art-directed hide, their hips under their skin, the meditative sideways metronomics of the jaw.

The fistulated—or "holey," as the students like to say—cow has been an ag-school standard for decades. My husband Ed recalls, as a child, hearing from his dad about the cow at Rutgers with "a window in its side." The operation is simple. The bottom of a coffee can is traced with chalk on the cow, a topical anesthetic applied, and the circle cut from the hide, along with a matching opening in the rumen. The two holes are stitched together and the hole is outfitted with a plastic stopper. It is little more barbaric than the earlobe plugs of my local Peet's barista or Ariel's facial adornments. "The animal rights people come out here expecting a glass window with a sash and sill," said DePeters. He handed me a protective plastic veterinary sleeve that extended to my shoulder and directed me to position myself to the side of the opening. When a fistulated cow coughs, if it has been eating, wet plant matter sometimes blows out of the hole.

DePeters took some photographs of me with my right arm in 101.5. The cow appears unmoved. I look like I've seen God. I was in all the way to my armpit and still could not reach the bottom of the rumen. I could feel strong, steady squeezes and movements, almost more industrial than biological. I felt like I'd stuck my arm into a fermentation vat with an automated mixing paddle at the bottom, and I basically had.

Ancient man was omnivorous—a scavenger as much as a predator. Often enough, his steak dinner was shared with millions of potentially harmful bacteria. Thus the human stomach, unlike the ruminant's, concerns itself with disinfection more than holding capacity. But even scavenged meals were sporadic, and some degree of storage was needed. How compliant is the human stomach? That depends on what you use it for.

# 10

## Stuffed

### THE SCIENCE OF EATING
### YOURSELF TO DEATH

O N APRIL 22, 1891, a fifty-two-year-old carriage driver in the city of Stockholm swallowed the contents of a bottle of prescription opium pills. Mr. L., as he became known, was found by his landlord and taken to a hospital, where the staff got busy with the tools of overdose: a funnel, a length of tubing, and luke-warm water to flush out and dilute the drug. The technique is known today as pumping the stomach, but in the case report it was referred to as gastric rinsing. The term gives a deceptive air of daintiness to the proceedings, as though Mr. L.'s stomach were a camisole in need of a little freshening. Hardly. The patient was slumped in a chair, thinly attached to his wits, while the medics loaded his stomach, multiple times in fast succession. With each filling, the organ appeared to hold more, which should have been a clue. Mr. L. had sprung a leak.

If you define *eating* as the mechanical act of putting something in your mouth and swallowing it, you could say that Mr. L., in consuming the pills, had eaten himself to death. Generally that is the only way to do it, to eat oneself to death. Bursting a stomach by overfilling is a nearly impossible feat, owing to a series of protective reflexes. When the stomach stretches past a certain point—to accommodate a holiday dinner or chugged beer or the efforts of Swedish medical personnel—stretch receptors in the stomach wall cue the brain. The brain, in turn, issues a statement that you are full and it is time to stop. It will also, around the same time, undertake a transient lower esophageal sphincter relaxation, or TLESR, or burp. The sphincter at the top of the stomach briefly relaxes, venting gas and restoring a measure of safety and relief.

Sterner measures may be needed. "A lot of people, myself included from time to time, eat way the hell past that point," says dyspepsia expert Mike Jones, a gastroenterologist and professor of medicine at Virginia Commonwealth University. "Maybe they're stress-eating. Or it's just: 'You know what, that's some damn fine key lime pie.'" The caution signs grow more obvious: pain, nausea, and the final I-warned-you-buster—regurgitation. A healthy stomach will up and empty itself well before it reaches the breaking point.

Unless for some reason it can't. In the case of Mr. L., the opium had interfered. The patient had "shown strong urges to vomit," wrote Algot Key-Åberg in a case report published in a German medical journal after Mr. L.'s autopsy was completed, but he had been unable to manage it. Key-Åberg was a professor of medicine at the local university and a very thorough man. I had hired a translator named Ingeborg to read Key-Åberg's paper aloud to me. The description of Mr. L.'s stomach and the ten parallel rupture wounds ran to two and a half pages. At some point Ingeborg looked up from the page. "So I guess the rinsing did not work out."

Mr. L.'s was the first stomach in Key-Åberg's experience to

have ruptured by overfilling. The case, he wrote, "stands on its own in the literature." Medicine needed to know about this so that future rinsers and pumpers could be alerted to the danger. Was it the volume of water or the force of its flow that mattered more? "In order to gain more clarity," Key-Åberg continued, "I needed to experiment with the stomach of a cadaver." Ingeborg made a small noise. "These experiments I conducted in large numbers." For much of the spring, unclaimed Stockholm corpses, thirty in all, were delivered to Key-Åberg's lab and maneuvered into chairs in a "half-seated position." Here one longs for some of that Key-Åberg zeal for detail. Was the position designed to mimic Mr. L.'s posture during the treatment, or did it simply reflect the difficulty of persuading a corpse to assume the upright profile of a dinner guest?

Key-Åberg found that if the stomach's emergency venting and emptying systems are out of commission—because the person is in a narcotic stupor, say, or dead—the organ will typically rupture at three to four liters, around a gallon. If you pour slowly, with less force, it may hold out for six or seven liters.

Very, very occasionally, the stomach of a live, fully conscious individual will give way. In 1929, *Annals of Surgery* published a review of cases of spontaneous rupture—stomachs that surrendered without forceful impact or underlying weakness. Here were fourteen people who managed, despite the body's emergency ditching system, to eat themselves to death. The riskiest item in these people's stomachs was often the last to go in: bicarbonate of soda (aka, baking soda, and the key ingredient in Alka-Seltzer). Bicarbonate of soda brings relief two ways: by neutralizing stomach acid and by creating gas, which forces the TLESR. (Less often, the stomach-inflating gas comes from actively fermenting food or drink. The *Annals* roundup includes a man killed by "much young beer full of yeast," and two deaths by sauerkraut.)

More recently, a pair of Miami-Dade County medical examiners reported the case of a thirty-one-year-old bulimic psychologist found seminude and fully dead on her kitchen floor, her abdomen greatly distended by two-plus gallons of poorly chewed hot dogs, broccoli, and breakfast cereal. The MEs found the body slumped against a cabinet, "surrounded by an abundance of various foodstuffs, broken soft drink bottles, a can opener and an empty grocery bag" and—"the coup de grace"—a partially empty box of baking soda, the poor man's Alka-Seltzer. In this case, the greatly ballooned stomach had not burst; rather, it killed her by shoving her diaphragm up into her lungs and asphyxiating her. The pair theorized that the gas could have forced one of the poorly chewed hotdogs up against the esophageal sphincter, at the top of the stomach, and held it there, preventing the woman from burping or vomiting.

By way of underscoring the impressive pressure produced by the chemical reaction of sodium bicarbonate and acid, I direct you to any of the myriad websites devoted to Alka-Seltzer rockets. Or, less playfully, the works of P. Murdfield, who, in 1926, ruptured the stomachs of fresh cadavers by pouring in a half gallon of weak hydrochloric acid and then adding a little sodium bicarbonate.

A safer road to relief is to drink a few sips of something carbonated. Or to swallow some air. People who swallow air chronically—aerophagia is the clinical term—are known among gastroenterologists, or one of them anyway, as "belchers." "You see a lot of belchers," says Mike Jones. "They do this hard swallow, where they're gulping air. It's like this nervous tic. Probably two-thirds of them are totally unaware that they're doing it. You watch them do it right in front of you, and they're going, 'Doc, I'm belching, and I can't understand it.'"

In addition to the social side effects, chronic belching splashes the esophagus with an excess of gastric acid, which sloshes out of the stomach along with the gas. If this happens too much or too often, the acid burns the esophagus. Now you have another reason to visit Dr. Jones: heartburn. How much acid exposure is "too much"? More than about an hour a day, according to research by David Metz, the University of Pennsylvania gastroenterologist we met in the previous chapter. That's the cumulative time each day that the normal esophagus is exposed to gastric acid. (People with gastric reflux spend far more time bathing their pipes in acid; in their case the sphincter may be leaky.)

One of the surgical treatments for chronic gastric reflux, called fundoplication, occasionally creates problems with belching. Now you really, really need to keep away from the bicarbonate of soda. "I know a case, this was fifteen years ago, where the man ate a huge meal and then took an inordinate amount of Alka-Seltzer." Jones made an exploding sound into the telephone.* "It was like that Monty Python sketch, the Wafer-Thin Mint, where the guy is gorging himself and finally he goes, 'I'll just have this one wafer-thin mint. . . .' "

$I$ F A WOMAN's abdomen is stretched so far that her belly button is inside out, it is usually safe to assume she is pregnant. The woman wheeled into the emergency room of the Royal Liverpool

---

* Though you do read case reports in which patients say they heard a bursting noise, the experience is more often described as a sensation, as in "a sensation of giving way." The "sudden explosion" recalled by a seventy-two-year-old woman following a meal of cold meat, tea, and eight cups of water was more likely something she felt, not heard. (The old eight-cups-of-water-a-day advice should possibly be qualified with the clause, "but not all at once.")

Hospital at 4 A.M. on an unspecified date in 1984 was an exception. She turned out to be carrying a meal. As dinners go, this was triplets: two pounds of kidneys, one and a third pounds of liver, a half pound of steak, two eggs, a pound of cheese, a half pound of mushrooms, two pounds of carrots, a head of cauliflower, two large slices of bread, ten peaches, four pears, two apples, four bananas, two pounds each of plums and grapes, and two glasses of milk. *Nineteen pounds* of food. Though her stomach eventually ruptured and she died of sepsis, the organ heroically held out for several hours. Likewise, recall the other bulimic—the model with the badly chewed hotdogs and broccoli. She died of asphyxia; the stomach never actually ruptured.

Clearly some stomachs hold more than a gallon.

The only human to have come close to the poundage record set by the Liverpudlian is Takeru Kobayashi, who consumed eighteen pounds of cow brains in an eating competition. Kobayashi had a fifteen-minute time limit. Presumably he'd have bested, or worsted, or wursted, nineteen pounds had the timer not gone off. Most food records are not measured in pounds, so it is hard to know how many others have come close. Ben Monson, for instance, consumed sixty-five Mexican flautas. Who knows what the freightage was on that. I never before noted the similarity between *flautas* and *flatus*, but I bet Ben Monson has.

Bulimic models and professional eaters are career bingers. They challenge the body's limits on a regular basis. Here is my question: Is the ability to eat to extremes a matter of practice, or are some stomachs—and I'm not saying anything here about my husband Ed—naturally more compliant?

In 2006, medical science took a look. David Metz observed the stomachs of a competitive eater—Tim Janus, then ranked number 3 on the circuit, under the name Eater X—and a six-foot-two, 210-pound control subject, while the men spent twelve minutes eating as many hotdogs as they could. A side of high-density barium enabled Metz to follow the wieners' progress via fluoroscope. Metz had a theory I hadn't considered: that prodigious eaters were people with faster-than-normal gastric emptying times. In other words, their stomachs might be making more room by quickly dumping food out the back door into the small intestine. The opposite turned out to be true. After two hours, Eater X's stomach had emptied only a fourth of what he'd eaten, whereas the control eater's stomach, more in keeping with a typical stomach, had cleared out three-fourths.

Somewhere into the seventh dog, the control eater reported to Metz that he would be sick if he ate another bite. His stomach, on the fluoroscope, was barely distended beyond its starting size. Eater X, by contrast, effortlessly consumed thirty-six hotdogs, taking them down in pairs. His stomach, on the fluoroscope, became "a massively distended, food-filled sac occupying most of the upper abdomen." He claimed to feel no pain or nausea. He didn't even feel full.

But the question remains: Are prodigious eaters born with a naturally compliant stomach, or do they alter the organ over years of incremental stretching—the digestive version of the tribal lip plate? Is the lack of discomfort there from the start, or does it come from habitually overriding the brain's signals? The implication, for the rest of us, being that the more you overeat, the more you overeat.

By happenstance, a friend of mine is acquainted with Erik Denmark—aka Erik the Red, ranked number 7 nationally—and offered to put us in touch. (The two had met on the set of dLifeTV, a show about living with diabetes. That a diabetic man holds the record for fry-bread consumption is yet one more mystery of professional eating.) I asked Denmark, Is the successful glutton born or built? Both, it seems. Denmark recalled visits to McDonald's as a child, where he would finish, by himself, the twenty-piece family box of Chicken McNuggets. But Metz had the impression, based on conversations with Eater X, that nature trumped nurture. "It's a structural thing," he told me. "At rest their stomachs are not much bigger, but their ability to receptively relax is unbelievable. The stomach just expands and expands and expands."

Though Denmark agrees with Metz that genes matter—as he puts it, "very few people could eat sixty hotdogs no matter how hard they worked at it"—he considers the inherently stretchy stomach merely the foundation, the starting point, for a career that requires daily practice and training. "I think," he told me, "that it has more to do with how much you're willing to push your body past the point that you would ever want to go." Despite his natural assets, Erik the Red did not hit the ground running. At his first competition, he put away just under three pounds to the winner's six pounds. (In relating the story, Denmark does not bother to mention what the food had been. It doesn't seem to matter. Flavor fatigue sets in after three to five minutes; beyond that point everything is more or less equally revolting.)*

I asked Denmark why the body's safety mechanisms, specifically regurgitation, don't kick in. In fact, they do. "This is going to sound gross," he said, "but you just, you know, like, swallow it down and keep eating." Major-league eating judges define regurgitation as the point at which food comes out, not up. "It's like a speed bump that you just go over. It's mental." Yes.

All competitive eaters follow a conditioning regimen. The

---

* With one exception. While the consumption record for many foods exceeds eight and even ten pounds, no one has ever been able to eat more than four pounds of fruit cake.

cheapest and least fattening training material is water. Denmark can water-load about two gallons at a sitting. When he began his career, he could barely get through one. As a point of reference—and warning—recall that one gallon was the point at which Key-Åberg's cadavers' stomachs began to rupture. Part of this training is psychological. In addition to stretching the stomach, water-loading gets the competitor accustomed to the feeling of being grotesquely full.

David Metz has a theory, yet untested, that water-loading could be used as a therapy for dyspeptics—people whose stomachs hurt after a meal, though they appear to be healthy. A 2007 study showed that dyspeptic patients report feeling full after drinking significantly less water, as compared with a control group of healthy, nondyspeptic volunteers. Could these people take a cue from professional eaters and gradually train themselves—by conditioning their stomachs—to comfortably hold more? "I think it would be a worthwhile project," Metz says.

Additional support for the incremental stretching theory comes from the other end of the eating spectrum—the starvation end. A surgeon-commander by the name of Markowski noted in a 1947 *British Medical Journal* paper that the stomachs of the World War II prisoners he treated were stretched from the large volumes of low-quality food they needed to eat to get enough calories and nutrients to survive. He surmised that the chronic stretching might weaken the organ, and that this explained why the men's stomachs sometimes ruptured after relatively small meals. Though if this were true, you'd expect to see stomach ruptures in major-league eaters, and you don't. I would have assumed that the prisoners' stomachs had shrunk, and that that was why they ruptured. I asked Metz about this. He dismissed the notion that people's stomachs shrink if they skip meals or cut way back on how much they consume. He says that when people say they feel full more quickly after eating less, it is because their tolerance for food is diminished; the feedback loops that are stimulated for hormone and enzyme production don't work as well.

Here is what surprises me: people with capacious stomachs are no more likely to be obese. A study in the journal *Obesity Surgery* reported no significant differences in the size of the stomachs of morbidly obese people as compared with non-obese control subjects. It is hormones and metabolism, calories consumed and calories burned, that determine one's weight, not holding capacity. Erik the Red insists he does not—outside of competitions—overeat, even though he never feels full. He points out that however much willpower it takes to stop eating when you're full, it takes far more to keep going (and going).

The biggest surprise of all is that the medical literature does not contain a single case report of stomach rupture among competitive eaters. Which brings us full circle to Mr. L. and my original point. By and large, it's not how much you eat that kills you, it's what you eat—especially, as we're about to see, when what you are eating is ten dozen latex bundles of cocaine.

# 11

# Up Theirs

## THE ALIMENTARY CANAL
## AS CRIMINAL ACCOMPLICE

Should circumstance prevent a man from carrying his cigarettes and cell phone in his pants pocket, the rectum provides a workable alternative. So workable that well over a thousand pounds of tobacco and hundreds of cell phones are rectally smuggled into California state prisons each year. The contraband allows incarcerated gang members and narcotics dealers to make business calls from behind bars (and to enjoy a smoke while doing so).

"This came in on Friday." Lieutenant Gene Parks is a contraband interdiction officer at Avenal State Prison. He is making reference to a clear plastic garbage bag two-thirds full with what appear to be but are not yams. They are plugs of Golden Leaf pipe tobacco, sheathed in latex and tapered at one end for ease of insertion, and not into pipes. The garbage bag is a "drop"—bulk contraband—that was hidden on the nearby chicken farm where two to three hundred Avenal inmates commute from the prison to work. Had Parks's team not gotten to the bag first, the plugs would have been "keistered" by convicts into the prison yard, two or three and occasionally six at a time, and then laid like the eggs the men spend their days with.

A fruity tobacco smell has leached through the plastic. The Investigative Services Unit smells like a tobacconist's shop. A one-pound bag of Golden Leaf tobacco retails for around $25. On the Avenal yard, an ounce sells for as much as $100, putting the yard value of that $25 bag at $1,600. The penalty, should you get caught, is mild––a temporary loss of visitor privileges. "We've disposed of, maybe, in the hundreds of thousands of these," says Parks. Lieutenant Parks has wide, voltaic blue eyes and a flat, imperturbable speaking manner. The combination makes him seem at once jaded and amazed.

Parks takes me into a storage room and shows me a bank of a dozen small square lockers, one for each month's contraband cell phones.

"All of these," I ask, "were . . ."

"Hooped?" Parks forms a circle with his thumb and forefinger. As in, through the hoop. More slang for rectally imported. "Not all. Some."

Parks takes two steps and reaches for another large plastic bag. "This is all chargers." Other bags and boxes contain batteries, earbuds, SIM cards. The slang for the rectum is "prison wallet," but it could be "Radio Shack." On the way here, I stopped in the office of a block captain who wanted to tell me about an inmate who was caught with two boxes of staples, a pencil sharpener, sharpener blades, and three jumbo binder rings in his rectum. He became known as "OD," for *Office Depot*. They never found out what he intended to do with the stuff.

• • •

$T$HE HOOPERS OF Avenal use the rectum for the basic purpose for which it evolved: storage. The nether distances of the gastrointestinal tract are a holding chamber for what remains of a meal once the intestine has absorbed what it can of the nutrients. Water is absorbed from the digesta as it travels along, and if all goes optimally, it leaves the body around the time it's reached a manageable water content: somewhere between 2 on the Bristol Stool Scale* ("sausage-shaped but lumpy") and 5 ("soft blobs with clearcut edges"). The lovely upshot is that one need only attend to the emptying once or twice a day.

If you'll allow it, a closer look at the process. Six to eight times a day, unbeknownst to your thinking, feeling self, a peristaltic muscle contraction called a mass movement squeezes the contents of the colon farther along. Eating reliably triggers this, via something called the gastrocolic reflex. The bigger the meal, the more vigorous the push. Any older detritus that had been parked outside the rectum now gets loaded inside. In with the new, out with the old. "It's a defensive reflex," explains William Whitehead,[†]

---

* Available in four languages, with minor modifications. The Portuguese edition, for instance, makes a distinction between the sausage of Types 2 and 3 (referred to as *linguiça*, a fatter German-style product) and that of Type 4, which is compared to *salsicha* (the more traditional wiener). The Bristol scale is, after all, a communication aid for physicians and patients. The more specific phrasing was undertaken "for better comprehension across Brazil."

† In a more perfect world, Whitehead would be a dermatologist, just as my gastroenterologist is Dr. Terdiman, and the author of the journal article "Gastrointestinal Gas" is J. Fardy, and the headquarters of the International Academy of Proctology was Flushing, New York.

co-director of the Center for Functional Gastrointestinal and Motility Disorders at the University of North Carolina. It prevents the colon from bursting.

When a load pushes against the rectum walls with sufficient pressure—as measured by stretch receptors—the defecation reflex is triggered. (You can trigger it prematurely by bearing down; this raises the pressure on the rectum walls to the requisite level.) The defecation reflex causes the rectal wall muscles to contract—that is, squeeze—at the same time the anal sphincter muscles relax. To the conscious mind this registers as urgency—somewhere between "Hello" and "Drop what you're doing." The larger or more liquid the load, the more pressing the urge and the tougher it is to hold back. Water will leak out a very small opening. As one gut expert put it, "Not even the sphincter of Hercules can hold back water." Take this to its end point and you have the simple saline enema— and an urgency that is not easily, if at all, overridden.

Though you can surely try. The defecation reflex has a manual override. Learning to employ that override is the essence of toilet training. Clenching the anal sphincter aborts the reflex and causes the urgency to fade—in most cases, long enough to pull off the highway or finish the aria and get to a toilet. (For patients who struggle to hold back the tide—sufferers of overbearing "postprandial urgency"—gastroenterologists recommend smaller, more frequent meals so that mass movements provoke a less intense onward push.)

Ahmed Shafik, the late, great chronicler of lower body reflexes, vividly demonstrated the defecation reflex in his lab at Cairo University. Volunteers were outfitted with devices to measure the squeeze pressure of both the rectum and the anus. A saline-filled balloon played the role of Turd. Filling the balloon with about a cup of water distended the rectum to the point where the reflex was triggered. The researchers could see on their instruments the sharp increase in rectal pressure—the squeeze—and the simultaneous drop in anal pressure—the letting go. "An urgent sensation was felt and the balloon was expelled to the exterior." *Ta-da!* When the subject was instructed to hold back, the rectum relaxed and "urgency disappearance" ensued. Mission aborted.

Setting aside the occasional interference of enemas, intestinal bugs, and Egyptian proctologists, adult humans are rarely at the mercy of their bowels. We need not soil our bloomers or drop our trousers and succumb there and then to the urge. Respect your equipment, people. The rectum and anus, working in concert, are a force for civilized human behavior.

And, occasionally, uncivilized behavior. Lieutenant Parks and his colleagues have called up some highlights of security camera footage from the visiting room. On the monitor, we watch a man palm an apricot-sized packet of something illegal that his wife has just slipped him, and then reach behind his back and deep into the seat of his pants, all *while playing a board game with his son.*

Based on the boxiness of the monitor we are viewing, Avenal's computer hardware does not appear to have been upgraded since the turn of the century. Budgets are lean. When I asked why the prison doesn't install a Body Orifice Security Scanner (a high-tech imaging chair that relieves guards of the distasteful tedium of bend-over-and-spread), Parks laughed. There isn't even money to reorder business cards. The prison was built for twenty-five hundred men, and now houses fifty-seven hundred. Everything, right down to the pink plastic flyswatter in Visiting Services, is broken or old or both. Meanwhile, the inmates are watching movies on smuggled smartphones.

The newer smartphones contain enough metal to set off the Avenal metal detectors, so they are hooped mainly by one inmate, a man with a hip replacement. His hip gains him a pass from the metal detector. "And we can't X-ray him without a court order or someone from medical saying that it's medically necessary," says Parks. The man hoops two or three phones at a time. The yard price on a smartphone is $1,500. "That guy is making a pile of money." Probably more than Lieutenant Gene Parks.

Three smartphones—or tobacco plugs—is a load far larger than the cup of water in Ahmed Shafik's balloon study. Given what I've learned about the physiology of the human rectum, it must be a tremendous struggle to keep it all in.

"That's something you can ask them yourself." Parks has arranged an interview.

ASIDE FROM A basketball backboard (I changed that from *hoop*, as a courtesy to you), and a few chairs set in a receding slice of shade, Yard 4 is bare. With rocks, someone has spelled out "4-YARD" in the rubbly parched dirt beside the gate. I think of inuksuit, the signposts that Arctic travelers build by piling stone slabs. In prison, as in the Arctic, you express yourself with the little you have at hand.

My escort from the Avenal Public Information Office, Ed Borla, calls to a guard to open the gate. A few inmates glance over as we cross the prison yard, but most ignore us. I am really, I think to myself, getting old.

Like all the yards at Avenal, this one has a row of amenities, each identified with a hand-painted red block-letter sign: GYM, LIBRARY, LAUNDRY, COUNSELOR, CHAPEL. It's like a tiny homegrown strip mall. I wait in one of the staff offices while Borla goes to find the man I'll be interviewing. I ask the staffer whose office it is whether he knows what my inmate is in for. He types the number on his computer keyboard and then turns the monitor toward me. The cursor blinks calmly beneath the word *MURDER*, just like that, in capital letters.

Before I have time to process this interesting piece of new information, the prisoner arrives in the hallway outside. I will call him Rodriguez, because I agreed not to disclose his real surname. Borla points to an empty office across the hall. "You guys will be in there." I glance down at my list of questions, which includes "Might hooping be a form of what the *Journal of Homosexuality* calls 'masked anal manipulation'?"

I explain myself as best I can. Rodriguez doesn't seem to find my line of inquiry to be freakish or surprising. As one of Parks's colleagues said earlier, of hooping, "It's a way of life." Rodriguez begins at the beginning, twenty-some years ago, in San Quentin. He belonged to a gang, and a leader of that gang approached him with an assignment. "I was told, 'Look, somebody is going to get stabbed in the—'"

I can't make out his last few words. ". . . in the arm?"

Rodriguez suppresses a smile. The very thought of a gang leader ordering an arm injury. "In the *yard.*"

Rodriguez doesn't project the personality that his rap sheet suggests. He is friendly, engaged. He looks you in the eyes. Smiles easily. Has beautiful teeth. You'd be happy to sit next to him on a long flight. You would never take him for a prisoner were it not for his pants, which say "PRISONER" in 200-point type down the length of one thigh. That's kind of a giveaway.

Rodriguez was ordered to smuggle—from work detail into the prison—four wrapped metal blades, a package twelve inches long and two inches fat. If he refused, he was told, one of the blades would be used on him. It was a harrowing experience, but he managed it. Since then, he has mainly hooped tobacco. "If you're going to go to the hole"—the other hole, solitary confinement—"you wrap up your tobacco, your lighter, matches . . ."* In the air, Rodriguez traces the outline of the smoking kit. It strikes me as far larger than one of Shafik's balloons. I explain rectal stretch receptors and the defecation reflex. "Are you always having to fight to hold it in?" I have an awareness that I must seem like an unusual person.

"Eeeh, yeah but . . ." Rodriguez looks at the ceiling, as though searching for the right phrasing, or beseeching God to intervene. "It finds its spot." In physiological terms, the defecation reflex has been aborted. After a certain number of aborts, the body gets the message and backs off for a while.

Gut motility experts will tell you that things happen to people who habitually abort the urge to go. Most are not smugglers. They're what gastroenterologist Mike Jones calls the "one more thing crowd." "They need to go, but they've got to do one more thing first." Or they are "bathroom-averse"; they're reluctant to use public restrooms because someone might hear or smell them, or because they're anxious about germs. By continually aborting the urge, these people may inadvertently train themselves to do the opposite of what nature intended. Their automatic response to

---

* Back in 2007, while researching a different book, I came across a journal article with a lengthy list of foreign bodies removed from rectums by emergency room personnel over the years. Most were predictably shaped: bottles, salamis, a plantain, and so on. One "collection"—as multiple holdings were referred to—stood out as uniquely nonsensical: spectacles, magazine, and tobacco pouch. Now I understand! The man had been packing for solitary.

"the urge"—even in the privacy of their home—is to tighten up. The medical term is paradoxical sphincter contraction. You're pushing on the door at the same time you're holding it shut. It's a common cause of chronic constipation.* And one that all the fiber in the world won't cure.

"You can figure out these folks really easily," Jones says. "You stick your finger in their rectum and you go, 'Okay, push,' and you feel them clamp down."

A group of German constipation researchers point out that "untoward conditions during the anorectal examination"—e.g., a stranger has his finger up there—can incite the anal sphincter to contract. Thus paradoxical sphincter contraction can be an artifact of diagnostic exams.† Though the authors acknowledge that for some patients, paradoxical sphincter contraction is assuredly the cause of their woes.

The medical staff at Avenal report that constipation is a common complaint.

---

* Biofeedback can help. The anal sphincter can be briefly wired such that tightening and relaxing causes a circle on a computer screen to constrict and widen. The patient is instructed to bear down while keeping the circle wide. The maker of that program has one for children, called the Egg Drop Game, wherein clenching and relaxing causes a basket to move back and forth to catch a falling egg. The website of the American Egg Board has a version of the Egg Drop Game that does not require an anus (or cloaca) to play, just a cursor.

† Especially if the exam entails defecography, which is pretty much what it sounds like. The patient is the star in an X-ray movie viewed by an audience of technicians, interns, and radiologist. "As close to pornography as medicine will come," says gastroenterologist Mike Jones. Worse, the patient is passing a barium-infused "synthetic stool" crafted from a paste of plasticine (or in simpler days, rolled oats) and introduced wrong-way into the rectum. For the constipated patient, notes Jones, it can be a real ordeal. "It's like, 'Dude, if I could do this, I wouldn't be here now.'"

• • •

THE ALIMENTARY CANAL is an accommodating criminal accomplice, but it has limits. The fuller the rectum and the longer you hold it back, the sooner the urge returns. Like a digital alarm clock, the more you ignore it, the bossier it gets. Twenty-four hours is about the limit for the average hooper. After that, Rodriguez says, "your brain just keeps telling you it wants to use the restroom." I picture Rodriguez's brain, desperate but polite, tapping him on the shoulder.

Swallowing contraband packets rather than hooping them buys the smuggler extra time. That's one reason swallowing is the preferred carrying technique of the international drug mule. Out of the 4,972 alimentary canal smugglers caught in Frankfurt and Paris airports between 1985 and 2002, only 312 had the goods packed in their rectum. Everyone else had swallowed it. Even on a ten-hour Bogotá–to–Los Angeles flight, swallowed packets typically don't reach the rectum by the time the plane lands. Mules are instructed not to eat anything during the flight. In this way they avoid triggering mass movements of the colon. (They may also take antidiarrheal drugs that shut down peristaltic contractions.) Thus even a cavity search of a suspected "swallower" may fail to produce any evidence.

Swallowers present a legal conundrum in that border detentions are required by law to be brief. Agents may detain a suspected smuggler only long enough to search luggage—checked, carry-on, and anatomical—and confirm or refute their suspicions. In a case that turned the lowly defecation reflex into a matter of Supreme Court deliberation, Bogotá resident Rosa Montoya de Hernandez was held for sixteen hours by customs agents in the

Los Angeles International Airport. A patdown and strip search had revealed a stiff abdomen—for Montoya de Hernandez's gastrointestinal tract was packed with eighty-eight bags of cocaine—and two pairs of plastic underpants lined with paper towels. She was given a choice: agree to an X-ray or sit in a room with a garbage bag–lined wastebasket and a female customs agent charged with, as they say at Avenal, "panning for gold."*

Montoya de Hernandez refused the X-ray. She sat curled up in a chair, leaning to one side and exhibiting, to quote Court of Appeals documents, symptoms consistent with "heroic efforts to resist the usual calls of nature."

Unfortunately for drug mules, the usual calls of nature are amplified by anxiety. Anxiety causes a mild contraction of the muscles of the rectum walls. This reduces the receptacle's volume, which means it takes less filling to activate the stretch receptors and confer ye olde sense of urgency. Rodriguez confirms this: "You have to relax. If you're nervous, your body clenches up." (Even mild anxiety has this effect. Using rectal balloons and regretful volunteers, motility researcher William Whitehead found that anxious people tend to have, on average, smaller rectal volumes.) In an episode of markedly high anxiety—giving a speech, say, or smuggling heroin—the effect can be dramatic. It's the last thing an "alimentary canal smuggler" needs. Mike Jones tells the story of a drug mule whose sphinc-

---

* Customs officers at Frankfurt Airport have it easier. Suspects are brought to the glass toilet, a specially designed commode with a separate tank for viewing and hands-free rinsing—kind of an amped-up version of the inspecting shelf on some German toilets. P.S.: The common assumption that the "trophy shelf" reflects a uniquely German fascination with excrement is weakened by the fact that older Polish, Dutch, Austrian, and Czech toilets also feature this design. I prefer the explanation that these are the sausage nations, and that prewar pork products caused regular outbreaks of intestinal worms.

ter surrendered on a flight into O'Hare. The man retrieved the packets from the airplane toilet and, rather than wash them off and reswallow them, stuffed them into the socks he had on—with predictable and life-changing results.

Montoya de Hernandez's lawyer tried, unsuccessfully, to argue that the plastic underpants and the eight recent passport stamps into and out of Miami and Los Angeles * did not constitute a clear indication that she was smuggling, and that her lengthy detention had been in violation of her Fourth Amendment rights. The U.S. Court of Appeals for the Ninth Circuit, however, reversed the conviction. And on it went, until Montoya de Hernandez and her stalwart anus made their way to the highest court of the land.† With Justices William Brennan and Thurgood Marshall dissenting, the Supreme Court reversed the Court of Appeals judgment.

---

* Other red flags for customs agents include the unique breath odor created by gastric acid dissolving latex, and airline passengers who don't eat. For years, Avianca cabin crew would take note of international passengers who refused meals, and report the names to customs personnel upon landing.

† Occasionally the justice system has no choice but to step right in it. In *State of Iowa v. Steven Landis*, an inmate was convicted of squirting a correctional officer with a feces-filled toothpaste tube, a violation of Iowa Code section 708.3B, "inmate assault—bodily fluids or secretions." Landis appealed, contending that without expert testimony or scientific analysis of the officer's soiled shirt, the court had failed to prove that the substance was in fact feces. The state's case had been based on eyewitness, or in this case nosewitness, testimony from other correctional officers. When asked how he knew it was feces, one officer had told the jury, "It was a brown substance with a very strong smell of feces." The appeals judge felt this was sufficient.

My thanks to Judge Colleen Weiland, who drew my attention to the case and did me the favor of forwarding a logistical question to the presiding judge, Judge Mary Ann—may it please the author—Brown. "It appeared," Brown replied, "that he liquefied the material and then dripped it or sucked it into the tube."

By refusing an X-ray and resisting "the call of nature," the Court concluded, Montoya de Hernandez was herself responsible for the duration and discomfort of her detention. The phrase "the call of nature" occurs so many times in the text of the case that I found myself applying a David Attenborough accent as I read.

*United States v. Montoya de Hernandez* set the precedent for the 1990 case of Delaney Abi Odofin, who spent twenty-four *days* in detention before passing the first of his narcotics-filled balloons. "An otherwise permissible border detention," the Justia.com summary concluded, "does not run afoul of the Fourth Amendment simply because a detainee's intestinal fortitude leads to an unexpectedly long period of detention."

How is such fortitude even possible? Why didn't Odofin's mass contractions seize the day? Why didn't his colon burst? Whitehead explained that the body has yet another rupture-preventing protective mechanism. A rectum that remains distended long enough will eventually trigger a slowing or even a shutdown of the production line, all the way upstream to the stomach if need be. Contractions of the colon and small intestine wane, and gastric emptying slows. This mechanism was documented in a 1990 study in which twelve students at the University of Munich were paid to hold back as long as they could. To see, one, whether and how long it's possible to suppress the urge, and, two, what happens when you do. The authors were impressed. "Volunteers succeeded in suppressing the urge to defecate to an amazing extent." Having just read Odofin's case, I wasn't all that amazed. Only three of the twelve made it to the fourth day.

The other thing the Munich researchers reported, and a mild *duh* here: the longer the material was held back, the harder and more pellet-like—the more scybalous—it became. Because as long as it sits in the tube, moisture will keep on being absorbed from it.

The harder and drier the waste gets, the tougher it is to eject. Holding it in causes constipation. The authors concluded their work with a word of advice for constipates (to use the exotic and rarely employed noun form): "Follow each call to the stool." Or, in the words of a British physician quoted in *Inner Hygiene,* James Whorton's excellent and scholarly* history of constipation, "Allow nothing short of fire or endangered life to induce you to resist . . . nature's alvine† call."

Constipation is the least of an alimentary canal smuggler's worries. About 6 percent of drug mules suffer bowel blockages‡ when packets logjam or the ends of the condoms become entangled. And there are overdoses. In the early days of alimentary canal smuggling, mules would wrap drugs in single condoms or fingers of rubber gloves, a thickness sometimes dissolved clear through after a few hours in gastric acid. Depending on the quality of the latex, the drugs would also leach through intact packaging. In more than half

---

* Seriously, published by Oxford University Press. But highly readable. So much so that the person who took *Inner Hygiene* out of the UC Berkeley library before me had read it on New Year's Eve. I know this because she'd left behind her bookmark—a receipt from a Pinole, California, In-N-Out Burger dated December, 30, 2010—and because every so often as I read, I'd come upon bits of glitter. Had she brought the book along to a party, ducking into a side room to read about rectal dilators and slanted toilets as the party swirled around her? Or had she brought it to bed with her at 2 A.M., glitter falling from her hair as she read? If you know this girl, tell her I like her style.

† Of or relating to the belly or intestines. With crushing disappointment, I learned that Dr. Gregory Alvine is an orthopedist. Staff at the oxymoronic Alvine Foot & Ankle Center did not respond to a request for comment.

‡ You would think the percentage would be higher, but in fact 80 to 90 percent of nondigestible objects that make it down the esophagus pass the rest of their journey without incident. If a man can swallow and pass a partial denture, a drug mule has little to worry about.

the reported cases of cocaine-swallowers spanning 1975 to 1981, the suspect died of overdose. (An antidote exists for heroin, but not cocaine.) Insult to injury: should you die on the job, you run the risk of your accomplices gutting your carcass to recover the drugs,* as happened to two of the ten dead Miami-Dade County, Florida, drug mules whose cases were covered in the *American Journal of Forensic Medicine and Pathology* paper "Fatal Heroin Body Packing."

At Avenal, drugs are typically hooped rather than swallowed. Parks's unit regularly intercepts illegal narcotics, as well as an evolving assortment of prescription drugs. (Wellbutrin, Xanax, Adderall, and Vicodin are snorted for various off-label recreational effects. The Rogaine that appeared in a recent over-the-fence drop appears to have been sought for its intended purpose.) Rodriguez has had cell mates who've opted to swallow. Two died of overdose. "One, he had like six months left. I go, 'Don't do it, man, you're too close to the house.'"

I ask Rodriguez how close he is to the house. Dumb question. Rodriguez is in for life. I had assumed the killing was gang-related, but it was over a girl. "It wasn't even my girl." Rodriguez rubs his thigh and looks away briefly, acknowledging something long past but still sharp. "I'm not the kid I was when I came in." That was twenty-seven years ago. "I'm starting to get white hairs, man. I'm starting to go bald." He lowers his head, to show me the bald spot or to register shame, I'm not sure which.

I don't know what to say. I like Rodriguez, but I don't like murder. "Dude," I finally manage. "Was that Rogaine yours?"

---

* Close to but not quite the most egregious indignity bestowed on a corpse by drug dealers. Smugglers have occasionally recruited the mute services of a corpse being repatriated for burial and stuffed the entire length of the dead man's GI tract. Heroin sausage.

• • •

HERE IS ANOTHER reason so many drug mules prefer to swallow contraband, despite the risk of an overdose. "The rectum is taboo across many of the regions where mules originate. In the Caribbean and Latin America, any use of the cavity is automatically associated with homosexuality, which can still lead to a fatal beating in many communities." This is from an e-mail from Mark Johnson, of the UK firm rather hazily known as TRMG, or The Risk Management Group.

The rectal taboo is equally strong among Islamic terrorists. Johnson's colleague Justin Crump, CEO of the London firm Sibylline, told me about the suicide bomber who tried to kill Saudi Deputy Interior Minister Muhammad bin Nayef in his home in Jidda in August 2009. Since little remained of the bomber's lower torso, the location of the explosives became an item of fizzy speculation among terrorists and counterterror experts. "All the jihadist websites were saying it was a swallowed device, that he had it in his stomach." Crump believes it was simply taped in place behind the bomber's scrotum.

"What was interesting," said Crump of the web postings, "was that there was a massive reluctance to say it could have been stuffed up his bottom." He recalls examining photographs of the bombing aftermath with a source of his, a former Al Qaeda militant. "He was saying, 'Oh, yeah, look at the way his arms came off. Definitely swallowed, definitely swallowed.' He was really keen to head off any notion that . . ." Here Crump himself seemed to trip over the taboo. ". . . To head off the other option."

No recorded instance exists of a suicide bomb being concealed inside a terrorist's digestive tract. Swallowing or hooping explosives, as opposed to wearing them in a vest, would reduce the destructive potential by a factor of five or ten, Crump says, because the bomber's body absorbs most of the blast. Bin Nayef was no more than a few feet away from an explosive the size of a grenade, but because the bomber was squatting on it, the target walked away without serious injury.

The only reason to smuggle a bomb inside one's body would be to get it through a strict security system, as exists in most airports. Crump says it's not worth the trouble; it's almost impossible to bring down a plane with a cache of explosives small enough to be alimentarily smuggled. A packet the size of a cocktail wiener is about the limit of what can be swallowed without undue travail. An accomplice could push the explosive material into the bomber's stomach in the form of a long thin tube, but the bomber would still need to swallow the timing device and somehow keep the digestive juices from rendering it inoperable.

Crump says a rectal bomb wouldn't bring down a plane either. "At most, you'd blow the seat apart." I showed him a Fox News piece that quoted unnamed explosives experts saying that a body bomb containing as little as five ounces of PETN could "blow a considerable hole" in an airline's skin, causing it to crash. "Total codswallop," said Crump. As fans of the TV program *MythBusters* know, even blowing out a window in flight won't create explosive decompression. The cabin will depressurize, but as long as the oxygen masks drop, people are likely to survive. "Remember that Southwest 737?" asks Crump. "The roof panel ripped partway off and they were fine. As long as you've got the pilots at the controls, and the plane's got wings and a tail, it will still fly."

Most suicide bombers don't achieve their goals via the explosives themselves. It's shrapnel that kills people. The typical marketplace suicide bomb is packed with nails and ball bearings—things you can't get past the airport metal detectors. To make a bomb that could bring down a plane, you'd need something that is, ounce for ounce, more explosive than TNT or C-4. Generally speaking, the more explosive the material, the more unstable it is. Trip and fall, or cough in the security line with a stomach full of TATP, and you may explode prematurely.

Materials found at Osama bin Laden's compound in Pakistan are said to have included a plan for surgically implanting a bomb in a terrorist's body—"in the love handles," according to an unnamed U.S. government source quoted on the *Daily Beast*. (Breast implants have also been tossed around as a possibility.) Crump has heard credible rumors of Al Qaeda physicians having tried out body implantation on animals. "But here again," Crump said, "there are a lot of issues. How to detonate it. How to keep the body from absorbing most of the blast." How to protect the explosives and the detonator from moisture.

This was comforting, but only for a moment. "Really, why bother with all that?" Crump said. "With a bit of prior observation, I can generally figure out a way to avoid going through a body scanner at most international airports."

THE PREFERENCE IN California prisons for rectal smuggling is a little surprising given the preponderance of Latinos and African Americans—two populations that are, taken as a whole, somewhat less comfortable with homosexuality. Prison, I'm guessing, is a place where extenuating circumstances erode the stigma that otherwise attaches to extracurricular uses of the rectum.

Rodriguez speaks freely about the situation in Avenal. Rather than antagonize gay inmates, he says, gang leaders tend to employ them. "We call them 'vaults.' If they're reliable, the homies will approach them—'Hey, check it out, you want to make some money?'"

Everyone else has to practice to get up to speed. Rodriguez recalls his "cherry" assignment—the blades—as extremely painful. He says gang underlings are made to practice. I picture muscular, tattooed men puttering around the cell with soap bars or salt shakers on board. Lieutenant Parks showed me an 8 × 10 photograph of what he said was a practice item, one that landed the apprentice in Medical Services. Deodorant sticks had been pushed into either end of a cardboard toilet paper tube and wrapped in tape. "As you can see," he said in his characteristic deadpan, "it's a rather large piece." (Rodriguez says it was hooped on a bet.)

"To avoid anal laceration, dilation may have to be performed progressively over a period of several weeks or months." This quote comes from a journal, but it is not a corrections industry journal or even an emergency medicine or proctology journal. It's from the *Journal of Homosexuality*. A corrections or even a proctology journal would not have gone on, in the very next sentence, to say, "Rowan and Gillette (1978) have described the case of a man who derived sexual pleasure from inflating his rectum with a bicycle tire pump." (As I did not pursue the reference, I remain ignorant of this man's fate and whether he exceeded the recommended PSI of the human rectum.)

Air and water (in the form of enemas) are the safest route to recreational distention because of the dependable ease of their removal. (An exception must be made for liquids that harden into solids. See "Rectal Impaction following Enema with Concrete Mix.") Solid objects tend to "get away from you," says gastroenterologist Mike Jones. "There's lubricant on the object, on the hands, you're in the throes of excitement and you're trying to grab it, and it's like, *gone.*" The ensuing panic makes it worse. Recall that anxiety causes clenching.

In the words of Anna Dhody, the ghoulishly ebullient Mütter Museum curator, "Every hospital has an ass box." The emergency medical literature is rife with case reports full of nouns you don't expect to see in a journal: *oil can, parsnip, cattle horn, umbrella handle.* The verb of choice, by the way, is *deliver.* As in: "This suction must be broken to deliver such glass containers." "A concrete cast of the rectum was delivered without incident."

One paper on the subject looked at thirty-five emergency room cases, all of them men. An explanation for the preponderance of males can be found in the aforementioned *Journal of Homosexuality* paper: "For males, dilation of the rectum . . . causes increasing pressure on the prostate gland and seminal vesicles, thus producing sensations that may be interpreted as sexual by some individuals." (The author, or perhaps there are two by the same name, appears to be a man of divergent interests. I found a list of his books on Goodreads.com. *Colorado above Treeline*, the list begins. *Life of a Soldier on the Western Frontier.* And then, nestled between *Medicine in the Old West* and *Exploring the Colorado High Country*, was *The Enema: A Textbook and Reference Manual.*)

Any discussion of the sexuality of the digestive tract must inevitably touch on the anus. Anal tissue is among the most densely innervated on the human body. It has to be. It requires a lot of information to do its job. The anus has to be able to tell what's knocking at its door: Is it solid, liquid, or gas? And then selectively release either all of it or one part of it. The consequences of a misread are dire. As Mike Jones put it, "You don't want to choose poorly." People who understand anatomy are often cowed by the feats of the lowly anus. "Think of it," said Robert Rosenbluth, a physician whose acquaintance I made at the start of this book. "No engineer could design something as multifunctional and fine-tuned as an anus. To call someone an asshole is really bragging him up."

The point I had been making is that nerve-rich tissue, regardless of its day-to-day function, tends to be an erogenous zone. Is it possible that these people who wind up in the emergency room are just folks whose anal play toys escaped into the interior?

Some, perhaps, but not all. Anal sensitivity cannot explain the man with the lemon and the cold cream jar. It cannot explain 402 stones. It cannot explain brachioproctic eroticism.* Research done by sexologist Thomas Lowry in the 1980s confirms the existence of a separate and devoted group of people whose specific joy derives from the sensation of stretching or filling. Lowry sent me a copy of his paper and the questionnaire he'd used to gather his data. Item 12 was a drawing of an arm, with the instructions, "Indicate with a line the deepest you have been penetrated." Suffice it to say that the anus, exquisitely sensitive though it may be, does not lie at the heart of these people's passions. Suffice it to say that some people enjoy Exploring the Colorado High Country.

Gustav Simon was the doctor for them. In 1873, Simon pioneered† the "high introduction" of a whole hand, "richly oiled,"

---

* A term coined by sexologist Thomas Lowry. In his efforts to research fisting, Lowry found himself writing letters to strangers at academic institutions that would begin like this: "Dear Dr. Brender: We spoke on the phone several months ago about 'fist-fucking.' At that time you mentioned two surgical articles." There was no academic term, so eventually Lowry made one up. "I Googled it recently," he told me, "and found over 2,000 hits. Made me chuckle."

† Simon refined his technique on cadavers, rupturing a bowel or two along the way, and then began offering training seminars. Cadavers were replaced with live, chloroformed women, thighs flexed on their abdomens. "A large number of professors and physicians" flew all the way to Heidelberg to practice "the forcible entrance."

into the rectum. This was done with the other hand pressed to the abdomen, to palpate the pelvic organs and check for abnormalities. (Gynecologists employ the method today, though typically hold themselves to two fingers.) Any resulting "pain in the parts," Simon assured the reader, was fleeting.

Mike Jones explains the arousal-by-stretching phenomenon by way of shared wiring. Defecation, orgasm, and arousal all fall under the purview of the sacral nerves. The massive vaginal stretch of childbirth sometimes produces orgasm, as can, at least in one diverting case study, defecation. Jeremy Agnew, in his 1985 paper "Some Anatomical and Physiological Aspects of Anal Sexual Practices," wrote, "Contraction of the anus upon manipulation of the clitoris during physical examination is often observed by gynecologists." Which kind of makes you wonder who Jeremy Agnew's gynecologist was.

I have a question, and forgive me in advance. If filling the rectum with stones or concrete or arms can be a direct flight to ecstasy, why is constipation so universally a misery? Or is it? Are there people who derive sexual gratification from self-manufactured filler? Is the urge to go ever complicated by the urge to come?

I accosted William Whitehead with these questions. "A lot of visceral sensation seems to follow what's been called a kind of Janus-faced function," he managed—meaning pleasure and pain on different sides of the same head. He had sidestepped the constipation question. Not wishing to be a pain in the parts, I lobbed the question over to Mike Jones's court.

"I think that the difference is that constipation is very rarely a self-determined event." What Jones was getting at, I believe, is that sexual arousal depends on the players and the circumstances. The difference between Ping-Pong balls and scybala is the difference between sexual intercourse and getting a Pap smear.

Most fans of back-door activities probably enjoy a combo plate of rectal and anal sensations. Why else would someone have invented the anal violin? Agnew describes this unusual item as an ivory ball with catgut attached. "The ball is inserted into the rectum while a partner strokes the attached string with a type of violin bow, thus transmitting vibrations to the anal sensory end organs," and puzzlement to the neighbors.

I never asked Rodriguez my question about "masked anal manipulation." (The term refers to gratification of anal carnality via seemingly nonsexual behaviors. It does not necessarily, though surely can, involve a Lone Ranger getup.) It seems to me no masking is needed: that men in prison can be fairly open about their anal intents. If a prisoner puts an iPhone up his rectum, it's because he wants to use it or sell it. If, on the other hand, he puts a toilet brush up there, he is seeking something more ineffable. Rodriguez told me about this one. "They took him out on a gurney, man. The *handle* was sticking out."

I told Rodriguez about the 402 stones.

"The rectum will stretch. Believe that."

THOUGH THERE HAS yet to be a case of a terrorist detonating a bomb in his alimentary canal, explosions inside the digestive tract are well documented. Flatus is mostly hydrogen, mixed with (in a third of us) methane. Both gases are flammable, a fact that occasionally becomes obvious in the endoscopy suite. As in volume 36 of the journal *Endoscopy*: "A loud explosion occurred in the colon immediately after the first spark induced by argon plasma coagulation." And again in volume 39: "Immediately on starting to treat the first of these angiodysplasias with APC, a loud gas explosion took place." And finally in *Gastrointestinal Endoscopy*, volume 67: "The authors reported that a loud gas explosion was heard during the treatment of the first of the angiodysplasias." Intestinal gas is not always funny.

Deuxième année. — N° 580.
Huit pages : CINQ centimes
Dimanche 18 Mars 1900

# Le Petit Parisien

## SUPPLÉMENT LITTÉRAIRE ILLUSTRÉ

### DIRECTION: 18, rue d'Enghien, PARIS

TOUS LES JOURS
Le Petit Parisien
5 CENTIMES

TOUS LES JEUDIS
SUPPLÉMENT LITTÉRAIRE
5 CENTIMES

TERRIBLE EXPLOSION A BORD DU PAQUEBOT « LA FRANCE »

# 12

## *Inflammable You*

### FUN WITH HYDROGEN AND METHANE

*L*ONG BEFORE ANYONE put a cautering wand up anyone else's patoot, the dangers of flammable* bowel gas were well known. If you let manure sit, as any farmer can tell you, bacteria will break it down into more elemental components. Some of these are of value to farmers as fertilizer, which they can pump from their manure pit out onto their crops.† Others—hydrogen,

---

* *Flammable* is a safety-conscious version of *inflammable*. In the 1920s, the National Fire Protection Association urged the change out of concern that people were interpreting the prefix *in* to mean "not"—as it does in *insane*. Though surely those same people must have wondered why it was necessary to warn of the presence of *gas that will not burst into flame.*

† "Work with your neighbors," urges the *Southeast Iowa Snouts & Tails Newsletter.* "Inquire about any outdoor events in the neighborhood such as weddings, cookouts and such to avoid manure application prior to those events." Unless your neighbors are also swine farmers, who apparently don't mind that sort of thing. The next item in the newsletter is a Manure Injection Field Demonstration "followed by a free lunch."

say, and methane—will blow the roof off the hog barn. Here is the Safe Farm Program channeling Beatrix Potter in a methane-safety radio spot: "It has no smell. It has no color. It often lurks about, but fails to leave a trace."

Methane and hydrogen are explosive in concentrations higher than 4 to 5 percent. The foam on liquid manure in pits is 60 percent methane. Farmers may know this, but their families sometimes don't. Which explains why the University of Minnesota Extension Service's farm safety curriculum includes instructions for a children's classroom Manure Pit Display. ("You will need: . . . toy cow, pig, and bull [1/32 scale], an aquarium, one pound of dry composted manure . . . and chocolate kisses . . . to simulate manure on top of floor [optional].")

Like a Manure Pit Display, the human colon is a scaled-down version of a biowaste storage tank. It is an anaerobic environment, meaning it provides the oxygen-free living that methane-producing bacteria need to thrive. It is packed with fermentable creature waste. As they do in manure pits, bacteria break down the waste in order to live off it, creating gaseous by-products in the process. Most voluminously, bacteria make hydrogen. Their gas becomes your gas. Up to 80 percent of flatus is hydrogen. About a third of us also harbor bacteria that produce methane—a key component in the "natural gas" supplied by utility companies. (At least two-thirds of us harbor a belief that methane producers' farts burn blue, like the pilot light on a gas stove. Sadly, a YouTube search unearthed no evidence.)

The flammability of methane and hydrogen is part of the reason for the seeming overkill of protracted bowel-cleansing that precedes a colonoscopy. When gastroenterologists find a polyp during a screening, they will usually remove it while they're in there, using a snare with an electrocoagulating option to staunch the bleeding. They do not want to worry about igniting a rogue pocket of combustible gas—as happened in France, in the summer of 1977, to fatal effect.

At a university hospital in Nancy, a sixty-nine-year-old man arrived at the Services des Maladies de l'Appareil Digestif (French for "Gastroenterology Department"). With the current set to 4, the doctor began a simple polypectomy. Eight seconds into it, an explosion was heard. "The patient jerked upwards off the endoscopy table," reads the case report, and the colonoscope was "completely ejected" (French for "launched from the rectum like a torpedo").

What was strange was that the Frenchman had followed his colonoscopy prep instructions to the letter. The culprit, in this case, had been the laxative. The staff had prescribed a solution of mannitol, a sugar alcohol similar to sorbitol, the likely laxative agent in prunes. Though the man's colon contained no fecal matter, it still contained bacteria, hungry bacteria that feasted on the mannitol and produced enough hydrogen to set the stage for an internal Hindenburg scenario. A study done five years later found potentially explosive concentrations of hydrogen or methane, or both, in six out of ten patients prepped with mannitol.

This is no excuse for putting off your colonoscopy. Mannitol is no longer used, and doctors routinely blow air or nonflammable carbon dioxide into the colon as they work, which dilutes any pockets of hydrogen or methane. (Inflating the colon also helps them see what they're doing. And creates the magnificent, billowing flatulence that rings through the colonoscopy recovery room.)

Outside the body, intestinal hydrogen and methane pose no danger. The act of passing flatus dilutes the gases, mixing them with the air in the room and lowering the concentration to levels well below combustibility. As anyone who has typed *pyroflatu-*

*lence* into YouTube is aware, the match would have to contact the gas the second it's blown from the body.

In the early days of the space program, NASA fretted about flammable astronaut flatus building up inside the tiny, hermetically sealed space capsules. A researcher presenting at the 1960s conference "Nutrition in Space and Related Waste Problems" was concerned enough to suggest that astronauts be selected from "that part of our population producing little or no methane or hydrogen." NASA used to keep flatus expert Michael Levitt—whom you'll meet shortly—on retainer as a consultant. Levitt assured them the capsules were large enough and the air inside sufficiently well circulated that intestinal contributions of hydrogen and methane were unlikely to reach a dangerous concentration. NASA was understandably wary. An earlier decision to circulate 100 percent oxygen in the capsules led to the deaths of all three Apollo 1 astronauts when a spark ignited a fierce fire during a launch-pad test.

EARLY ONE MORNING in the winter of 1890, a young British factory worker leaned from his bed to check the time. It was not yet dawn, the streets of Manchester still dark and shuttered. As he struck a match to see the hands of his timepiece, he happened to emit a belch. "To his consternation," wrote Dr. James McNaught in the *British Medical Journal*, "the gas took fire, burned his face and lips considerably, and set fire to his moustache."

Cases of "inflammable eructation"—McNaught cites eight other cases—are perplexing. The gas in a typical belch is either carbon dioxide (from carbonated drinks) or air swallowed while eating or drinking, both nonflammable. The healthy human stomach, unlike the colon, does not produce hydrogen or methane. Gastric acid's job is to kill microorganisms; without them there can be no hydrogen- or methane-producing fermentation. Even if a relative few bacteria survive in a stomach—and some species can, we now know—the chymified food is passed on to the small intestine too quickly for fermentation to make much headway.

McNaught reached for his stomach tube. It had been five hours since the factory worker ate, a time lag normally sufficient for the stomach to finish its duties and pass the chyme along to the small intestine. Yet up came a pint and a half of a sour-smelling, soupy matter with a sediment of "grumous* remains of food." And gas, lots of it, visible as a head of frothy bubbles, foaming and bursting like the contents of a mad scientist's beaker.

To identify the gas and confirm its flammability, McNaught had only to collect some from the headspace of the beaker and set it alight. But that's no fun. Instead, on a different day, McNaught had the man yet again visit his office. Through a tube, he poured water into the misbehaving stomach, to displace the gas. As he did so, he held a flame to the invisible plume issuing from the man's mouth. "The result . . . was to produce a flame of dimensions alarming to both the patient and myself." Maybe I'm projecting, but a poorly suppressed schoolboy glee occasionally surfaces in McNaught's writing, setting it off from the typical Hippocratic benevolence of *British Medical Journal* prose. If I had a medical license, I fear I'd be a Dr. McNaught.

It turned out that owing to strictures of the pylorus, the stom-

---

* Meaning "clotted or lumpy." *Grumous* is one of many evocative words that deserve to break free from medical dictionaries and join the ranks of day-to-day vocabulary. Likewise, *glabrous* ("smooth and hairless"), *periblepsis* ("the wild look of delirium"), and *maculate* ("spotted").

ach's lower sphincter,* the food in the young man's stomach was held back an uncommonly long time. Plus, McNaught claimed to have cultivated strains of acid-resistant, gas-producing bacteria. Carbohydrates plus bacteria plus time and body heat equals fermentation.

The story made me curious about cows. As we learned earlier, the rumen is a vast fermentation pit, a massive bacterial slum. A grazing cow can produce a hundred gallons of methane a day, vented, as stomach gases typically are, through the mouth. You would think that cow-belch-lighting would rival cow-tipping as a late-night diversion for bored rural youth. How is it that growing up in New Hampshire I never heard a cow belch? My ag pal Ed DePeters had the answer. When a ruminant is feeling bloated and needs to make room in her rumen, she pushes out some methane, but instead of belching it up, she can shift her internal tubing to reroute the gas down into the lungs and then quietly exhale it. To, say, a pronghorn out on the savannah, quiet can be key to survival. "Ungulates in the wild tend to go off and hide someplace while they ruminate," DePeters explained. "If a lion walks by and hears a loud *urrp* . . ." *Sayonara*, antelope.

Because my readers, perhaps more than anyone else's, might be inspired to head out to the pasture with a lighter in their pocket and bovine malfeasance in their heart, let me add this: lighting a cow's breath will not produce a McNaughtian geyser of flame. Because of the afore-described methane rerouting system, the gas is diluted by nonflammable gases in the breath. For ignition, you would need the sort of concentrated blast that is a belch. And cows don't belch.

---

* *Pylorus* is Greek for "gatekeeper." That's all. As you were.

Snakes don't either, but they can, under certain circumstances, create an inflammable eructation of literally mythical proportions. For this story, we leave Ed DePeters in his muck boots and feed cap and turn to our snake digestion man in Alabama, Stephen Secor. First, a little background: Many plant-eating animals lack rumens, so some fermenting takes place in the cecum, an anatomical pouch at the junction of the small intestine and the colon. These same plant-eaters—horses, rabbits, koalas, to name three—tend to have a larger-than-average cecum. Pythons and boas do too, which struck Secor as odd, because they're carnivores. Why, he wondered, would a meat-eater need a vegetation digestion unit? Secor theorized that perhaps these snakes had evolved ceca as a way to digest and take advantage of plant matter inside the stomachs of their prey.

To test his theory, Secor fed rats* to some of the pythons in his lab at the University of Alabama and hooked them up to a gas chromatograph. He tracked the hydrogen level in their exhalations as they digested whole rats over the course of four days. He did see a spike, but it appeared long before the rat arrived at the python's cecum. Instead, Secor suspected, the hydrogen spikes were the result of the decomposing, gas-bloated rat bursting inside the python. "One thing led to another." (Secor's way of saying he popped a bloated rat corpse and measured the hydrogen

---

* Purchased in bulk from RodentPro.com. Life is cheap at RodentPro, as cheap as sixteen cents for an extra small pinky (a one-day-old frozen feeder mouse). Mice are also available in fuzzy (ten to fifteen days old), white peach fuzzy ("Just the right size when a pinky is too small and a fuzzy is too large"), hopper, weanling, and adult. Feeder rats and guinea pigs are sized like T-shirts: XS, S, M, L, XL, and XXL. RodentPro gift certificates are available. Because nothing says "I love you" like $100 of dead rodents delivered to the doorstep.

that came off it.) Suspicion confirmed. The hydrogen level was "through the roof." Secor had stumbled onto a biological explanation for the myth of the fire-breathing dragon. Stay with me. This is very cool.

Roll the calendar back a few millennia and picture yourself in a hairy outfit, dragging home a python you have hunted. *Hunted* is maybe the wrong word. The python was digesting a whole gazelle and was in no condition to fight or flee. You rounded a bend and there it was, Neanderthal turducken. Gazython. The fact that the gazelle is partially decomposed does not bother you. Early man was a scavenger as well as a hunter. He was used to stinking meat. And those decomp gases are key to our story. Which I now turn over to Secor.

"So this python is full of gas. You set it down by the campfire because you're going to eat it. Somebody kicks it or steps on it, and all this hydrogen shoots out of its mouth." Hydrogen, as the you and I of today know but the you and I of the Pleistocene did not know, starts to be flammable at a concentration of 4 percent. And hydrogen, as Stephen Secor showed, comes out of a decomposing animal at a concentration of about 10 percent. Secor made a flamethrowery *vhooosh* sound. "There's your fire-breathing serpent. Imagine the stories that would generate. Over a couple thousand years, you've got yourself a legend." He did some digging. The oldest stories of fire-breathing dragons come from Africa and south China: where the giant snakes are.

# 13

# Dead Man's Bloat

## AND OTHER DIVERTING TALES FROM THE HISTORY OF FLATULENCE RESEARCH

*T*HE SAME QUALITY that has allowed Mylar to rival latex as the material of choice for party balloons has secured its place in modern-day flatus research. Mylar is airtight. Your helium-filled Mylar Get Well Soon balloon will continue to float long after you are discharged from the hospital. The Mylar balloon I inflated in 1995 as part of a flatulence study might still, had anyone kept it, contain gas I produced by eating two-thirds of a pound of chili in the Kligerman Regional Digestive Disease Center cafeteria.

Alan Kligerman is the *Kligerman* of the Kligerman center, and he is the *Ak* in AkPharma, the company that founded the center and created Beano.* The active ingredient in Beano is an enzyme

---

* AkPharma has since sold the Beano brand to pharma giant GlaxoSmith-Kline. As part of a marketing campaign, GSK's website included an online University of Gas. Hoping to matriculate—or at least buy a sweatshirt—I clicked on the video. The stately campus building in the background was

that breaks down certain complex carbohydrates, called oligosac-charides, found in large quantities in beans and other legumes. You have this enzyme in your colon, courtesy of bacteria that live there. Because your small intestine can't absorb these complex carbohydrates, they carry on into your colon, where bacteria and their enzymes break them down—and create a lot of hydrogen in the process. Translation: beans make people gassy. Adding Beano to chili while it's still on the table preempts this. It's like having a surrogate predigest your beans.

I had visited Kligerman's lab for a magazine piece. I still have my notes and interview transcripts, and a teal Beano windbreaker* that Kligerman gave me, but the details are hazy. I recall eating my carefully weighed chili at a table with Kligerman and Betty Corson, the voice of the Beano Hotline. My notes attest that a man called Len was also there. My lunch mates were eating the chili too, though they weren't part of the study. They were just people who like beans, or had come to like them, since AkPharma purchased them in volume and cans could usually be found in the cupboards of the employee kitchen.

"I'll open up a can of black beans, and I'll eat the whole can," said Betty.

Len was nodding. "I'll take a can of baked beans. Pour the liquid off. That's what I'll eat for lunch a lot of times. I hate to

---

instantly familiar to me as Baker Library at Dartmouth College, where my parents once worked. Given what I know of the Dartmouth frat scene, it was kind of apt, but I ratted GSK out anyway. The president's office did not seem to share my outrage ("At this time, I don't have a comment from President Kim about the Beano University of Gas"), but a cease-and-desist letter eventually went out, and the image was removed.

* But one example of the sly marketing genius of AkPharma. Beano was also the sponsor of a team of hot air balloonists in a prominent race.

admit it, but I'm one of the 50 percent of Americans that's not troubled by beans."

When someone at AkPharma says, "Troubled by beans," *trouble* doesn't refer to the embarrassment caused by the sounds or smells of flatulence. (Hydrogen and methane are odorless, remember.) *Trouble* refers to the pain and discomfort caused by gas inflating your colon. When the colon balloons, it activates stretch receptors that send a message to your brain, which your brain forwards to you as pain. Like most pain, it's an alarm, a warning system. Because stretching can be a prelude to bursting, your brain is highly motivated to let you know what's happening down there.

The older you get, the slacker the muscles of your colon become and the more easily the organ balloons. As Len gaily remarked, "We get flabbier all over, inside and out." Sixty percent of Beano customers, he said, are over fifty-five. People with coronary artery disease whose doctors steer them away from fats and red meat are often advised to incorporate beans into their diet as a replacement protein. "Some of these people," said Kligerman, "would come back to their doctors, going, 'I'll take my chances on a second heart attack over all this gas.'" Cardiologists in the fat-fearing 1980s handed out Beano sample bottles like Halloween candy.

The other food group that troubles the middle-aged gut is dairy. About 75 percent of Asians, African Americans, and Native Americans are deficient in lactase, an enzyme secreted in the small intestine that breaks down lactose, a sugar found in milk products. In Caucasians the rate is around 25 percent. Most can digest milk sugars while they're young, but they begin to lose the ability as they age. "Once you're beyond suckling age," Kligerman pointed out, "there's no biological reason for you to absorb lactose." Were it not for the persistent hand of the dairy lobby—Got Marketing?—the notion of grown-ups drinking milk by the glass might seem as odd here as it does in much of the rest of the world.

Milk products follow the same biological plot line as beans. An ornery carbohydrate makes it to the colon intact because the small intestine couldn't break it down into something absorbable. Colon bacteria go to town on it, spewing clouds of hydrogen in the process. Gastroenterologists can easily diagnose malabsorption of lactose (or gluten, for that matter). In the Bay Area, where I live, people prefer to self-diagnose. And misdiagnose. "Dairy sugar often travels with dairy fat, and big fat loads are hard on the gut," says gastroenterologist Mike Jones. "People who claim lactose intolerance tend to also voice a belief that they're gluten-intolerant. Usually with no evidence of either."

True lactose malabsorption is no picnic. This was the source of the prodigious flatus of the pseudonymous patient A. O. Sutalf,* documented in 1974 and reported in the august pages of the *New England Journal of Medicine*. Mr. Sutalf, his identity a closely guarded secret to this day, passed gas an average of thirty-four times a day. By comparison, the lactose-tolerating adult will toot on average no more than twenty-two times a day,† peaking twice: five hours after lunch and five hours after dinner. Len maintained that the 5 P.M. peak was at least in part man-made: "You've been holding it in at work, and as soon as you get in the car to drive home you let it out."

---

\* ?ti teG

† I brought Levitt a scrap of notebook paper covered with hash marks, the score card of an anonymous family member who kept track for two days, totaling thirty-five and thirty-nine. "Yeah," Levitt said, "every time I give a talk someone comes up and tells me twenty-two is way too low."

Whereupon Kligerman frowned. Earlier, when Len tried to tell a story that began, "There was a guy on my floor freshman year . . . ," Kligerman threw this particular bucket of cold water: "This is not a humorous subject."

When Kligerman got up to take a phone call, I scooted my chair over to Betty Corson's side. I wanted to know who'd been calling the Beano Hotline lately. She told me about a woman whose boyfriend kept pulling over to "check the air in the tires." More typically, it is women, mostly of my mother's generation, who don't want anyone, ever, under any circumstances, to hear them. Like the gassy nun at the Holy Spirit–Corpus Christi Monastery, who had called earlier in the day. "She talked very quietly," said Corson.

Why not just avoid legumes? Some people can't, said Corson. I challenged her to provide a single instance of a human being forced to eat beans. She came back with "refried-bean tasters." They exist and they have called the hotline. "Can you imagine?" She slapped the table. "Honest to God." With Kligerman gone from the table, the conversation had loosened its tie a little.

I know one other example of beans eaten against one's will. Inmates in solitary confinement in state prisons are sometimes fed a single, nutritiously complete but wholly unappetizing food called Nutraloaf. (Often these are convicts who've attacked someone with their silverware. Nutraloaf is an entire meal you can pick up and eat by hand.) Beans are invariably a main ingredient, as are bread crumbs, whole wheat flour, and cabbage: major gas generators all. Inmates in several states have sued on the grounds that Nutraloaf three times a day constitutes cruel and unusual punishment. In the article I read, taste was the issue, but an elderly convict could probably build a case out of gas pain.

When Kligerman returned, he was carrying what looked like a potato chip bag with a snorkel apparatus at one end. He explained that he needed to get a baseline reading before I ate my beans. He handed the device to me. "When you do your blows—"

It was unlike Kligerman to employ slang in reference to flatus. It quickly became clear that the snorkelly thing, like any snorkelly thing, went in the mouth, not the rectum. I was both relieved and disappointed: he was doing a breath hydrogen test. If you know the amount of hydrogen someone is exhaling orally, it's a simple matter to extrapolate the amount they're exhaling rectally. This is because a fixed percentage of the hydrogen produced in the colon is absorbed into the blood and, when it reaches the lungs, exhaled. The breath hydrogen test has given flatus researchers a simple, consistent measure of gas production that does not require the subject to fart into a balloon.

Up through the 1970s, however, that's how it was done. A retired bean scientist told me the story of a flatus research project carried out by the extremely appropriately named Colin Leakey, at a food science facility in Chipping Campden, near Stratford-upon-Avon. If I'd been a tourist passing through, I might have skipped the Shakespeare and gone over to Chipping Campden to have a peek. "People walked around in gowns"—hospital, presumably, not ballroom—"with a tube coming down and around and up into a balloon." Stateside, in 1941, J. M. Beazell and A. C. Ivy rigged up something similar: "The gas was collected in a thick-walled rubber balloon by means of a 22 French colon tube which was inserted into the rectum about 10 cm [4 inches]. In order to hold the tube in place, a broad strip of dental rubber dam was attached to it at the point where it emerged from the rectum and this was brought up snugly along the gluteal fold and fixed to the abdomen and back by means of adhesive tape. With this arrangement the subjects were able to remain ambulatory with surprisingly little discomfort."

The researchers were fooling themselves, says Michael Levitt. "The rectal tube is . . . uncomfortable, tends to plug, and cannot be used for prolonged periods in free-living subjects," he wrote in a 1996 paper. For studies of gas volume, he preferred the "flatographic recording" technique, wherein the subjects would make a notation in a special diary of each "passage." The method isn't entirely dependable, however, because different people's passages may contain widely varying amounts of gas, depending on whether . . . how shall I put this? On whether they are my husband or my mother-in-law. On whether they expel it with gusto or try to hold it in, letting it out in many tiny squeakers and falsely upping their flatographic tally.

Len pointed out a related shortcoming to the breath hydrogen test. When people, stereotypically women, hold in their gas, they absorb more of it into their bloodstream, "so it comes out in the breath." This artificially raises their breath hydrogen numbers and may serve to explain the occasional highly counterintuitive finding that women are more flatulent than men.

"Right, Alan?"

Kligerman stirred his chili. "I don't know, Len. I don't know what the ultimate fate of a suppressed fart is."


RECTAL TUBES AND breath hydrogen bags have their drawbacks, but either is an improvement on the original methodology. One of the earliest flatus studies on record was carried out by the Parisian physician François Magendie. In 1816, Magendie published a paper entitled "Note on the Intestinal Gas of a Healthy Man." The title is misleading, for although the man in question suffered no illness, he was dead and missing his head. "In Paris," Magendie wrote in *Annales de Chimie et de Physique*, "the condemned ordinarily, one hour or two before their execution, have a light meal." With red wine. So French! "Digestion is thus fully active at the moment of their death." From 1814 to 1815, the city fathers of Paris, seeming also to have lost their heads, agreed to release the bodies of four guillotined men to Magendie's lab for a study on the chemical makeup of flatus. One to four hours after the blade had dropped, Magendie extracted gas from four points along the digestive tract and measured what he could.

One of the prisoners Magendie "opened" had consumed lentils as part of his last meal. I would have expected this man to have had the highest hydrogen level—legumes, as we just learned, being the largest supplier of unabsorbed carbohydrates to hungry colon bacteria. Oddly, the highest hydrogen level came out of the prisoner who had eaten *"pain de prison et du fromage de Gruyère."* Gruyère cheese and "prison bread." Were Paris jailers serving some sort of early French precursor to Nutraloaf? Probably not. For many people, unabsorbed carbohydrates from wheat are a sizable contributor to gas. And if you're going to be dead in two hours, there's no reason not to fill up on bread.

What amazed me about Magendie, aside from his zest for gore, was this: using instruments available in 1814, he was able to detect hydrogen sulfide, a gas that typically makes up one-ten-thousandth of the gas produced in the human colon. It's possible the instrument Magendie used was, in fact, his nose. The human olfactory system detects the rotten-egg smell of hydrogen sulfide at the practically nonexistent rate of .02 parts per million. Though present in no more than trace amounts, hydrogen sulfide is, in the words of Michael Levitt, "the most important determinant of flatus odor." He would know.

# 14

# *Smelling a Rat*

## DOES NOXIOUS FLATUS DO
## MORE THAN CLEAR A ROOM?

**M**ICHAEL LEVITT DID not set out to make his mark on the world by parsing the secrets of noxious flatus. His fellowship advisor had the idea. The gas chromatograph had just come into use as a laboratory tool, and no one had yet had the ingenuity—or nerve—to apply the technology to human emissions. "He called me into his office," Levitt recalls. "He said, 'I think you ought to study gas.' I said, 'Why's that?' He said, 'Because you're pretty much of an incompetent, and this way if you discover anything, at least it'll be new, and you'll be able to publish something.'"

Levitt published thirty-four papers on flatus. He identified the three sulfur gases responsible for flatus odor. He showed that it is mainly trapped methane gas, not dietary fiber or fat, that makes the floater float. Most memorably, to this mind anyway, he invented the flatus-trapping Mylar "pantaloon."

"Even now," he says of his flatus work, "it overshadows everything else I do." Levitt and I are sitting in a conference room upstairs from his lab at the Minneapolis VA Medical Center. Levitt has a goofy, lopsided smile and a pale complexion. I couldn't recall, while writing this, whether his hair was gray, so I typed his identifiers into Google Images. A photograph of a can of baked beans came up.

For the record, here are some of Michael Levitt's other contributions to medicine: He invented the breath hydrogen test, which originated not as a flatulence assessment technique but to diagnose malabsorption of carbohydrates in the small intestine. He debunked a diet fad for foods made with "nonabsorbable" carbohydrates. He showed that the wriggling movements of the villi are the key to intestinal stirring and to healthy absorption of nutrients. "I wrote the book on intestinal stirring."

After what I judge to be a sufficient number of follow-up questions on intestinal stirring, I ask whether it might be possible to see the Mylar pantaloons.

Levitt designed the garment for a pair of studies that aimed both to identify the gases responsible for noxious flatus and to test devices claiming to adsorb—the formal term for binding to something's surface—those gases. He doesn't know where they are stored, but digs out a photograph of a woman standing in the lab, modeling them. Shown uninflated, they fit more snugly than I'd pictured them. The material is silver, crinkly, and reflective. They're the sort of clothes baked potatoes wear.

I ask Levitt whether it was difficult to recruit volunteers for the flatus studies. It wasn't, partly because the subjects were paid for their contributions. People who sell their flatus are more or less the same crowd who turn up to sell their blood.

"What was hard," Levitt says, "was finding the judges." Levitt needed a pair of odor judges to take "several sniffs" and rate the noxiousness—from "no odor" to "very offensive"—of each of the sixteen people's flatal contributions.* The hypothesis was that noxiousness would correlate with the combined concentrations of the three sulfur gases. And it did.

Curious as to which olfactory notes the different sulfur gases contributed to the overall bouquet of flatus, Levitt purchased samples of the three gases from a chemical supply house. The judges agreed on the following descriptors: "rotten eggs" for hydrogen sulfide, the gas with the strongest correlation to stink; "decomposing vegetables" for methanethiol; and "sweet" for dimethyl sulfide. Though lesser players contribute as well, it is for the most part these three notes, in subtly shifting combinations and percentages, that create the infinite olfactory variety of human flatus. To quote Alan Kligerman, "A gas smell is as characteristic of a person as a fingerprint is." But harder to dust for.

The great variety of flatus smells—from person to person and from meal to meal—presented a quandary for the second phase of the study, the evaluation of various odor-eliminating products. Which—whose—wind should represent the average American's? No one's, as it turned out. Using mean amounts from chromatograph readouts as his recipe and commercially synthesized gases as the raw ingredients, Levitt concocted a lab mixture deemed by the judges "to have a distinctly objectionable odour resembling that of flatus." He reverse-engineered a fart. This "artificial flatus" was put to work testing a variety of activated-charcoal products: underwear, adhesive-backed underwear pads,

---

* It could be worse. In a study of malodorous dog flatulence carried out at the Waltham Centre for Pet Nutrition in Leicestershire, England, the far end point of the scale was "unbearable odor."

and chair cushions. (Activated charcoal is known to be effective at binding sulfur gases. The circulating air supply in NASA spacesuits is filtered with activated charcoal, lest astronauts' flatus be blown across their face three times a minute for the remainder of the spacewalk.)

In a separate study to simulate real-life gas-passing conditions, Levitt taped a tube beside the subject's anus, beneath the charcoal pad or underpant and the subject's pants. (Cushions were strapped in place.) The subject then pulled the Mylar pantaloons over whatever product was being tested, and an assistant duct-taped the cuffs and waistband to the skin. Levitt hit a switch, and just under a half cup (100 milliliters) of synthesized flatus shot through the tube for two seconds—Levitt's best guess for the size and life span of a typical fart. "Immediately following gas instillation," wrote Levitt in the final paper, "air inside the pantaloons was constantly mixed via vigorous palpation over a 30-second period." Levitt claims to have no video footage. Last, a syringe was fitted into a port in the Mylar to withdraw the gas, and Levitt measured the sulfur gases the charcoal had failed to trap.

The challenge, it turned out, lies in bringing the gas fully into contact with the charcoal—easy with an airtight spacesuit, less so a business suit. Seat cushions were relatively useless, most products trapping a scant 20 percent of the sulfur gases. The underwear pads delivered a 55 to 77 percent reduction, their efficacy compromised by "rectal gas blow-by": the tendency of the wind to glance off the pad and out the sides rather than penetrate it. The seventy-dollar briefs performed best, adsorbing virtually all sulfur gases, though it was unclear how many wearings they were good for. And given the cost, in terms of both cash and self-esteem, they would seem to have a limited market.

. . .

$A$S AN ALTERNATIVE to wearing activated charcoal or gluing it to your underpants, you could swallow some pills. But don't bother, because Levitt has done a study on this too. Activated charcoal pills did not "appreciably influence the liberation of fecal gases." Levitt surmised that the binding sites were saturated by the time the charcoal made it to the rectum.

Bismuth pills, on the other hand—and Levitt has tested these too—reduce 100 percent of sulfur gas odor. Bismuth is the *bism* in Pepto-Bismol. Daily doses of Pepto-Bismol can irritate the gut, but not bismuth subgallate, the active ingredient in Devrom "internal deodorant" pills.

I had never before heard of Devrom. This may be because mainstream magazines often refuse to run the company's ads.* Devrom's president, Jason Mihalopoulos, e-mailed me a full-page ad he had hoped to run in *Reader's Digest* and *AARP* magazine. A smiling gray-haired couple stand arm in arm below the boldface headline "Smelly Flatulence? Not since we started using Devrom!" Mihalopoulos was told he could not use the phrases *smelly flatulence* and *stinky odor* or the word *stool*. One of the magazines suggested changing the copy to say that the product "eliminates intestinal gas," but that's not what Devrom does. That's what Beano does. So unless you read the *Journal of Wound Ostomy &*

---

* The exception being the *Saturday Evening Post*. The *Post* has a robust tolerance for graphic medical copy, as evidenced by the November 2011 article "Lumps and Bumps on Your Pet: What Could They Be?"

*Continence Nursing** or the *International Journal of Obesity Surgery,* you won't see the happy, internally deodorized Devrom couple.

The noxious-rectal-gas taboo in mainstream advertising has proved stronger and more lasting than that of condoms and even vibrators, which now turn up in brazenly suggestive ads on cable television (though still under the century-old euphemism "massager"). Mihalopoulos told me the editors of a CNBC feature on quirky businesses refused to air a segment about Parthenon, the family-run business that makes Devrom. "People don't like to hear *flatulence*," he said, quickly adding that he meant the word. Or anyway, people think people don't.

Given the obvious strength of the taboo, I wondered who had posed for the Devrom ad. How much do you have to pay people to appear in a full-page ad in a national magazine, talking about their smelly flatulence?

"Oh, I'd be shocked if someone would be willing to pose in an ad we'd run," Mihalopoulos said. "It's a stock photo." Meaning anyone, for a fee, can run the image for whatever purpose they choose. The couple probably have no idea. Think twice before you sign a model release form.[†]

Most Devrom customers are people with extenuating digestive circumstances. They've had their stomachs stapled or bypassed to shed weight, or they've had all or most of a diseased gut

---

* These nurses deserve a special award that is difficult to picture.

† Back in the 1980s when everyone looked a bit off, my friend Tim and his brothers had some publicity shots taken of their band. Eventually the photographer sold the rights to a stock photo agency. Years later, one of the images turned up on a greeting card. The inside said, "Greetings from the Dork Club."

removed and they're excreting into an ostomy pouch. Mihalo-poulos explained that, depending on how high up the opening is, the pouch may need to be emptied every few hours. Less time in the colon means less water is absorbed. The runnier the waste, the more surface area is exposed to the air and the more volatiles escape to reach the nose. "If you were to use the restroom at the airport, say . . ." Mihalopoulos paused to figure out where he was going with this. "You could tell right off that someone was emptying their pouch."

It seemed, then, that we were not even talking about passing gas. "No, that too," said Mihalopoulos. He explained that some people with an ostomy pouch will open a corner of it to let a little gas out. "It's like Tupperware."*

Mihalopoulos didn't have data he could share regarding the number of people who were taking Devrom to defang garden-variety flatus odor, rather than because of a medical situation. I'm guessing there aren't very many of them, and I think I know why. I think I know what's keeping internal deodorant from charging ahead as a mainstream product. I'm going to let Beano inventor Alan Kligerman tell you what it is. "When I talk to people," he told me, "when I really get them down to the nitty-gritty, I don't know anybody, really, in their heart of hearts, who has any objection to the smell of their own." And, unlike bad breath or stinky

---

* Before you try to tell me that the proper verb for degassing Tupperware is *burp* not *fart*, let me pass along the words of a Tupperware spokeswoman I interviewed in 1998: "We don't say *burp* anymore. Now we talk about making the seal 'whisper.'" I don't think *whisper* is a good substitute for *burp*, but it makes a lovely, poetic euphemism for the silent rectal passage. *Forsooth, Horatio, even her whispers beguile me.*

feet, "smelly flatulence" is everyone's problem.* And thus really, no one's.

As with the first bottle of Scope, the first bottle of Devrom, Mihalopoulos confirmed, is often left anonymously by a coworker or purchased by a spouse. "They themselves don't object to it," he said, "it" referring to the smell, not the purchase. Levitt said he is constantly approached at cocktail parties by women complaining about their husbands' gas. He has never once heard a husband complain about a wife, despite this scientifically proven (by Levitt) fact: "the flatus of women has a significantly greater concentration of hydrogen sulfide and was deemed to have a significantly worse odour by both judges." (However, this is likely balanced out by the male's "greater volume of gas per passage.")

The Devrom company is to be commended for not aggressively pushing internal deodorant on the public at large. Good for you, Jason Mihalopoulos, for not following in the springtime-fresh footsteps of douche marketers and, most recently, the Fleet enema company.[†] "Keep your backcountry clean," says the Fleet Naturals ad copy, over an image of pristine mountain wilderness. "Created specifically for rectal cleansing . . . Mild enough for daily use." *Really?* On top of gargling, on top of powdering our feet and

---

* Though some more than others, depending on your flora. Some people have more of the sulfur-producing bacteria. The sulfur-spewers, by the way, prefer to colonize the descending colon, the part nearest the rectum. This is why noxious flatus tends to have heat. The composting happens right near the exit, so the flatus is, as gastroenterologist Mike Jones put it, "hot off the press."

† Inventors of the world's first purgative superhero, EneMan: an enema bottle with arms and legs and a pointy nozzle head, dressed in a green cape. (Plush-toy EneMen occasionally turn up on eBay, not that I was looking.)

perfuming our armpits, now we should worry that our assholes smell?

I later stumbled upon a "Tell Your Patients . . ." press release that Fleet had sent out to physicians. (One of them had posted it on his blog.) It turns out that Fleet Naturals is a product "for before or after anal intimacy." Well okay then.

The simplest strategy for bouts of noxious flatus is to not care. Or perhaps to take the advice of a gastroenterologist I know: get a dog. (To blame.) Barring that, a person might try to steer clear of certain foods,* the ones that provide bacteria with the raw materials for making sulfur compounds. The main offender is red meat.† Cruciferous vegetables (broccoli, cabbage, brussels sprouts, cauliflower) can also kick up a stink. As can garlic, dried and sulfured fruit (for example, apricots), certain aromatic spices, and, for reasons unclear, beer. In short, so many delightful things that a sane person would, I like to think, rather have the gas.

---

* So strongly does stink depend on diet that the gases emanating from a rehydrated 6,400-year-old turd have been used to reconstruct the diet of an ancient "defecator." Or so claimed J. G. Moore and colleagues in the 1984 article "Fecal Odorgrams." The title refers to a method of analyzing waste fumes via a gas chromatograph and a "sniffing port." Nowadays diet can be determined by sequencing the DNA of the food in fossilized turds, so no one need ever create (or send) a Fecal Odorgram.

† Decomposing protein stinks: "aged" cheese, rotten eggs, corpses, dead skin on the bottoms of your feet. "Morning breath" is hydrogen sulfide released by bacteria consuming shed tongue cells while you mouth-breathe for eight hours; saliva normally washes the debris away. The stench is a warning: this item contains a lot of bacteria and could (depending on which bacteria they are) make you sick. The scariest, stinkiest cuisines are in countries where both food and refrigeration are scarce. Rural Sudanese eat fermented (that is, decomposing) caterpillar, frog, and, less proteinaceously, heifer urine. Yet one more reason tourism has been slow to catch on in the Sudan.

• • •

*I* TRAVELED TO MINNESOTA with a fantasy that Michael Levitt might be able to whip up a batch of artificial flatus. I'm curious to see how close Science can get to Nature. Levitt smiles one of those placeholder smiles that buy you a moment to phrase your no. He elects to fob me off on his research partner Julie Furne, who has the ingredients downstairs in the lab. I recognize Furne's name from the pantaloon studies. It turns out she had been one of the odor judges.

Things haven't changed all that much for Julie Furne. We find her in the lab, syringing gas out of a plastic vial in which a raisin-sized rat turd has been incubating at ninety-nine degrees. (She and Levitt are investigating the relationship between intestinal hydrogen sulfide and colitis. More on this shortly.)

Furne recently arrived at her fifth decade, her brown hair beginning to silver at the hairline but a girlish humor still intact. Instead of a lab coat, she wears a muted orange heather cardigan, vintage from the fifties, I'm guessing. There was probably a time when you could have pressed this sweater to your face and smelled traces of hair spray or homemade pot roast. Probably you wouldn't have that experience now.

"This is Mary," says Levitt. "She'd like to sniff some gases. But don't kill her."

Hydrogen sulfide is as lethal, molecule for molecule, as cyanide. This may explain why humans evolved such exquisite sensitivity to its smell. Repellent odors are unpleasant but often helpful in terms of not dying. As with any poison, dosage makes the killer. The concentration of hydrogen sulfide in offensive human flatus is around 1 to 3 parts per million. Harmless. Ramp it up to 1,000 parts per million—as can exist in manure pits and sewage tanks—and a couple breaths can cause respiratory paralysis and suffocation. Workers die this way often enough that a pair of physicians, writing in a medical journal, coined a name for it: dung lung.* Hydrogen sulfide is so swiftly lethal that farm- and workplace-safety organizations urge anyone who enters a manure pit or attempts to clear a blocked sewage pipe to wear a self-contained breathing apparatus. Which may explain the man my husband and I once saw walking along a sidewalk in San Francisco in a wet suit with a toilet plunger over his shoulder. "Hell of a clog," Ed said.

It is fitting that the Devil is said to smell of sulfur. Hydrogen sulfide is a diabolical killer. Its telltale rotten-egg smell, screamingly obvious at 10 parts per million, disappears at concentrations above 150 parts per million; the olfactory nerves become paralyzed. Without the odor to warn them, coworkers and family members may rush into a manure pit to rescue the fallen. Whole families have been taken out in a catastrophic "chain of death." One case report included a police photo taken after the victims had been pulled from the mire and laid on the ground. It was a wrenching play on the family portrait, the four adults arranged in a row in matching knee-high muck boots, black bars over their eyes. The farmer had gone in to unclog a pipe. Both he and the worker who tried to rescue him collapsed and died. The farmer's mother found the two, hurried down the ladder and also succumbed. Then the son came along. And on it went, all the way to a team of pathologists nearly overcome in a poorly ventilated autopsy room.

Hydrogen sulfide is a reliable way to kill oneself—as well as the

---

* One of the physicians was a Dr. Crapo, who would, you'd think, have long ago ceased to find that sort of thing amusing.

people who try to save you. In 80 percent of the hydrogen sulfide suicides in this country, emergency personnel or good Samaritans have been sickened by the fumes when they tried to help. One suicide in Japan caused the evacuation of 350 neighbors.

"Ask Julie how she felt," says Levitt over his shoulder as he leaves. The day of her training as a flatus odor judge, Furne worried that she'd poisoned herself. She was "sick as a dog" and had a headache all that evening. Vegetarian activist John Harvey Kellogg wrote that he had "known vigorous young men" who suffered "violent attacks" of headache from working in a lab with "the bowel discharges of a meat-eater."

The hydrogen sulfide pulled from the tube that held the fermenting feces of rat E2 clocks in at a concentration of 1,000 parts per million. "You don't want to smell that straight up," says Furne. She glances off to the side, reading out an imaginary headline: "AUTHOR KILLED BY FECAL ODOR." Furne has a homey northern Midwest accent, the voice of *Fargo*'s Margie, diluted to a nonlethal concentration.

But this is hydrogen sulfide off-gassing in a vial smaller than a lipstick. Are there circumstances in which ordinary concentrations can harm you? Are flatulent people a public health risk? The author of *Inner Hygiene*, James Whorton, quoted a nineteenth-century physician who thought so. He admonished the flatulent to hold in their gas for the sake of family and friends, saying, "It is as much a crime to poison a neighbor with gas, as with a poison more tangible." I wondered if there might be a grain of truth to this—inside a confined space, say. When it's cold, I tell Furne, I sometimes sleep with my head under the covers. Winter is brussels-sprout season, and they're Ed's favorite side dish.

Furne assures me there's enough air under a comforter to dilute a spouse's hydrogen sulfide and render it harmless. When I followed up with Levitt by e-mail, he concurred that the "passive inhaler" has no cause to worry.

Especially compared with the perpetrator. He who dealt it incurs a "relatively enormous exposure to hydrogen sulfide via absorption through the colonic mucosa." Or as John Harvey Kellogg rather more excitably put it, "If the mere breathing of the greatly diluted volatile poisons arising from such putrescent matter will produce highly unpleasant effects, how much more grave must be the effects when through the retention within the body . . . all of their poisonous contents are absorbed and sucked up into the blood and circulated throughout the body?" Levitt had been quick to add that no research has shown that absorption—of hydrogen sulfide or any other colon-residing breakdown component—into the bloodstream is harmful.

In matters of health, however, the public rarely requires proof. Most people trust intuition more than they trust studies. And the theory behind fecal self-poisoning—aka autointoxication—makes strong intuitive sense. "[People] reasoned that if feces are foul, then the body must be in the best condition when freest from such material," wrote Walter Alvarez in his wise and tide-turning 1919 essay in the *Journal of the American Medical Association*. The less time "feculent" poisons reside in our colons, the thinking went, the less we absorb into our blood, and the healthier we'll be. Autointoxication was one of the most pervasive and enduring concepts in the long, bloated history of medical pseudoscience.

As a diagnosis, a health buzzword, autointoxication peaked in the early 1900s. It was a natural offshoot of "miasma" theory. From the early to the late 1800s, before physicians figured out the role of microorganisms and insects in causing and spreading disease, much of the blame was placed on clouds of nonspecific toxic gases—or miasmas—emanating from open sewer flows, garbage dumps, even graves.

If one bought into the dangers of miasmas, it wasn't much of a leap to buy into the dangers of one's own internal sewage. Purveyors of laxatives and enema devices played up the connection, referring to the colon as "the human privy," "an obstructed sewer," "this cesspool of death and contagion." Whorton's book reproduces a magazine ad for the French laxative Jubol, showing tiny uniformed men on their hands and knees with scrub brushes and buckets inside a colon, like workers in the Paris sewers.*

It made no difference that neither the specific poisons nor the mechanisms by which they might be causing harm were known or named. In the realm of quackery, vague is better. "It met a need," wrote Whorton, "that medicine has felt in every age, providing an explanation and diagnosis for all those exasperating patients who insist they are sick, but are unable to present the physician with any clear organic pathology to prove it." Autointoxication was the gluten of the early 1900s.

Bogus diagnoses beget bogus cures. Around the turn of the last century, hosing the colon was big business, far bigger than it is today, and nowhere bigger than at 134 West Sixty-Fifth Street, home of Tyrrell's Hygienic Institute, a three-story New York

---

* Heartlessly, Jubol failed to provide its imaginary workers with tiny face masks. Or shoes! They're barefoot in there! In reality, it's people inside French sewers who deserve our concern, not people inside sewers inside French people. France's Department of Occupational Epidemiology found elevated rates of liver cancer among Parisian sewer workers, though most of them also drink to excess, and who can blame them.

brownstone dedicated to the manufacture and flatulent hyping of the J.B.L. Cascade colonic irrigator. *J.B.L.* stood for "Joy Beauty Life," suggesting that your $12.30 was purchasing something loftier than a nozzle-topped whoopee cushion.

"The Internal Bath is taken by sitting on the J.B.L. Cascade," states Charles Tyrrell in the 1936 promotional pamphlet *Why We Should Bathe Internally*. Tyrrell's prior business had been in rubber medical goods. Aside from the rectal nozzle protruding from its flank, the Cascade looked little different from one of Tyrrell's old water bottles.

Between businesses, Tyrrell had dabbled in small-press publishing. The experience served him well. He printed up thousands of thinly disguised promotional booklets that he distributed to pharmacists to hand out to patients. The gospel of autointoxication and internal putrescence was laid on thick and spiked with testimonials: from customers, doctors,\* clergy,† all wordily pro-

---

\* Most of them dead, bought, or similarly corrupt—like the purveyor of Medicine for the Prevention of Motherhood and (perhaps the fallback nostrum) Remedies for Children.

† Judging by the number of testimonials from priests, prelates, sisters, and superiors, religious celibates were avid embracers of rectal irrigation. Inside the J.B.L. Cascade files in the Historical Health Fraud collection of the American Medical Association archives, I found a "Dear Reverend Father" come-on—a special offer "being made to the Catholic Clergy only." Though Presbyterians found their way to it too; a satisfied Reverend J. H. M. wrote to say that he had "worn out" three bags over the years.

Balancing out testimonials from the gentle and the frocked was one by the trainer of the New York Giants from 1930 to 1932, Leonard Knowles. Knowles hinted but did not outright state that the players' regimen had included sessions with the Joy-Beauty-Life Cascade. In an unusual display of restraint, Charles Tyrrell did not take credit for the Giants' second- and third-place finishes in the National League during the time Knowles was with the team.

fessing their satisfaction and gratitude. Gone was their insomnia, their fatigue, their melancholia. Here was the fix for acne, bad breath, for lack of appetite and "loss of vim and snap." An internal bath would rid you of irritability, "outrageous cantankerousness," "the inability to hold down a job of lumber grading for over six months without quitting or getting fired." One set of before and after photos seemed to imply that a high colonic could transform an unkempt, drooping moustache into a vigorous, curlicued handlebar.

It seemed there was no medical condition so dire that an internal bath would not fix it. Mr. H. J. Wells of 342 Lincoln Avenue, Detroit, credited the Cascade with relieving his wife of "an accumulation of effete mucous tissue . . . in strips about half an inch wide and from four to six inches long." Mrs. Cora Ewing of Long Beach, California, waved good-bye to "a sack of pus above the left ovary." People thanked Tyrrell for curing their asthma, their rheumatism, their typhoid fever, and their jaundice. Paralysis even! Epilepsy! The medical claims were sufficiently far-fetched that Tyrrell felt a need to point out that the "disorders may be due to factors other than . . . autointoxication."

The American Medical Association's Bureau of Investigation received so many letters from outraged physicians that it drafted a form letter to send in response. "We plan to get around to this institution after a while," it promised. The first such letter in the Tyrrell Hygienic Institute file at the AMA archives is dated 1894, and the last, 1931, suggesting that a little more vim and snap might have been applied.

One member rose on his own to the task. In 1922, physician and autointoxication doubter Arthur Donaldson artificially and incontrovertibly constipated three dogs by temporarily sewing shut their anus. After four days, all the while eating regular meals of meat, milk, and bread, the dogs showed no physical symptoms beyond a mild loss of appetite—nothing to suggest a poisoning from within. All three, impressively, "seemed to be in fair spirits."

Donaldson didn't rest his case there. He withdrew small amounts of blood from his surgically constipated charges, once at the end of fifty-five hours, again at seventy-two hours, and finally at ninety-six hours. This he injected into the bloodstream of two normal, unconstipated dogs,* to see whether symptoms suggestive of "fecal poisoning" would develop. They did not.

Donaldson contended that the symptoms people and doctors were so quick to blame on autointoxication were in fact caused by the simple mechanics of constipation: rectal distention and irritation. To test the theory, he packed four men with turd-sized wads of cotton. After three hours, the men began to exhibit the sorts of symptoms commonly blamed on autointoxication. The moment the wads were removed, relief ensued. If fecal blood poisoning had been the culprit, relief would have taken far longer. It takes several hours for the liver and kidneys to clear chemicals from the system. The reek of asparagus pee, Walter Alvarez pointed out,

---

* As autointoxication experiments go, this one presents a comparatively minor affront to animal welfare. Less mildly, here is Frenchman Charles Bouchard, in 1893, referring to his laboratory rabbits: "I have practiced intravenous injection with the extracts of fecal matter. It produces depression and diarrhea." Which begs the question: If you are a caged lab animal under the care of a man who is liable, on any given day, to inject you with human excretions, is it possible to *be* any more depressed? Ask the animals over in Christian Herter's lab. Over the course of several months in 1907, Dr. Herter injected rabbits and guinea pigs with fecal extract from lions, tigers, wolves, elephants, camels, goats, buffalo, and horses. Herter wanted to see whether the shit he got from carnivores was more pernicious than the shit he got from herbivores. The rodents died either way, leading one to wonder about the shit he got from the humane society.

though not in those exact words, doesn't abate the moment you set down your fork. It lingers through the following morning. The very swiftness with which the enema brings relief itself refutes the premise of autointoxication.

In the incomparable phrasings of gastroenterologist Mike Jones, "Everybody who's bound up feels a whole lot better after a big dump. From where I sit, you don't need to invoke anything else."

THE ALTERNATIVE APPROACH to ridding the body of "faeculent poisons" was to eat so much fiber that digesta sped through the colon too quickly to generate them. Insoluble dietary fiber, or roughage, is the indigestible, nonfermentable parts of plants— internal yard trimmings that the gut cannot break down. This fiber sponges up water, contributing dramatically to fecal "bulk." The bulkier the trash, the sooner you need to empty the bin.

John Harvey Kellogg was the archbishop of roughage. The healthy colon, he maintained, empties itself three or four times a day. This was "Nature's Plan." As evidence, he cited the estimable bowel frequency of "wild animals, wild men, . . . infants and idiots." Kellogg's sources included the staff at "well-managed idiot asylums" and ape keepers at the London Zoo. Kellogg paid several visits to the latter "for the express purpose" of discussing the toilet habits of their charges. The chimpanzees, noted Kellogg, "move their bowels four to six times daily." All the more to throw at zoo visitors. Kellogg effected a habit of dressing in immaculate white suits, but probably not on the second and third visits.

Kellogg didn't gather data on the regularity of "wild men," but someone else sure did. In the early 1970s, epidemiologist A. R. P. Walker held a post at the South African Institute for Medical Research, affording easy access to Bantu people and others "pursuing a primitive manner of life." In his travels through South African villages, Walker noted that "unformed stools are frequently encountered among rural Bantu." One man's ruined footwear is another's eureka moment. The Bantu, Walker knew, were almost never diagnosed with Western digestive diseases. Was it because they ate so much fiber? Did their woody digesta exit the colon too quickly to inflict harm?

Walker got busy clocking stool: British versus Bantu. Subjects swallowed radio-opaque pellets and then "voided" into plastic bags that they labeled with the date and time. The bags were X-rayed* so researchers could see exactly how long it took the pellets to complete their journey. As with foot races, so with digestion: the slowest third of the Bantu were quicker than the fastest third of the Caucasians. This was because, Walker assumed, the Bantu ate a shitload of insoluble fiber in the form of millet and corn porridge.

Walker was the man behind bran. Papers published by him and, more recently, his research partner Denis Burkitt, fueled a decade-long fiber craze. Americans were forcing down unprecedented amounts of bran muffins, oatmeal, and high-fiber break-

---

* As an aside, Walker noted that "stools can be sieved to retrieve the pellets, thus avoiding the need for X-rays." Who would sieve when they could X-ray? Someone who long ago wore out his welcome in the radiology department. Based on the following, I'm guessing Walker may also have been pushing his luck with Bantu villagers. "Eighty to 98 percent of rural Bantu children," he marveled, can "produce a stool on request."

fast cereals. Whorton cited a 1984 survey that found a third of Americans eating more fiber to stay healthy.

You don't hear so much about fiber these days. Curious, I ran a PubMed search on cancer and dietary fiber. The most recent study, published in the *American Journal of Epidemiology* in 2010, followed three thousand Dutch men for thirteen years. Get a load: "Frequent bowel movements were associated with an increased risk of rectal cancer in men, and constipation was associated with a decreased risk." Mike Jones wasn't surprised. The medical community was never completely on board Burkitt's fiber train. "He was comparing the Bantu to, like, British naval recruits, guys who ate practically no fiber and they all smoked." Many other factors also set the British apart from rural black Africans—how do you control for them all? "It was correlation, not causality, and you really couldn't take it any further."

So why did we hear so much about fiber back then? Because, Jones said, there was money to be made: "things to go out and buy and eat more of." Walker and Burkitt wrote the tune, but it was the cereal companies that kept on playing it. Jones said that when he sat down and looked at the studies on dietary factors and colon cancer, the thing that stood out as a determinant of risk wasn't how much fiber you ate, but how many calories. The fewer calories, the lower the risk. No easy profits there.

$A$ND GET THIS. The newest research suggests that slower transit time—that is, longer exposure to your nasty stuff, may in fact be of benefit. Hydrogen sulfide appears to prevent inflammation and its sometime consequences, ulcerative colitis and cancer. In rodent studies, anyway, the gas has a significant anti-inflammatory effect on the walls of the digestive tract: the opposite of what aspirin does in there. Aspirin and ibuprofen combat inflammation everywhere *but* the stomach and bowel; there they *create* inflammation. Used in tandem with hydrogen sulfide, says Ken Olson, a professor of physiology at Indiana University School of Medicine and author of multiple papers on the topic, aspirin or ibuprofen may be thousands of times as potent at preventing tumor growth—at least, in mice and in laboratory-grown tumor cells. Human trials have not yet begun.

Hydrogen sulfide is not the devil. Beneath the danger and stench is a molecule as basic and indispensable as sodium chloride. The gas is produced in all of the body's tissues, all the time, regardless of what was for dinner. (Some recent thinking disagrees.) "It's a gasotransmitter, a signaling molecule, it has tremendous therapeutic potential," says Olson. "This is the hottest area in biomedicine right now."

The moral of the story is this: It takes an ill-advised mix of ignorance, arrogance, and profit motive to dismiss the wisdom of the human body in favor of some random notion you've hatched or heard and branded as true. By *wisdom* I mean the collective improvements of millions of years of evolution. The mind objects strongly to shit, but the body has no idea what we're on about.

HERE'S THE OTHER hitch with autointoxication. Absorbing things is primarily the business of the small intestine, not the colon. That's what the smaller tube, with its millions of villi, is for: delivering nutrients to the blood. The autointoxication zealots would counter that, as John Harvey Kellogg put it, "the foul fecal matters in the colon pass back into the small intestine." But, in fact, they don't. The ileocecal valve, the anatomical portal between the small intestine and the colon, opens in one direction only.

It is possible to force open the ileocecal valve from the wrong direction, but it does not happen naturally, in the course of day-to-day living. It has tended to happen unnaturally, while dead, on a slab in a nineteenth-century anatomy amphitheater with one end of a flexible tube disappearing up the rectum and the other attached to a pump. No less than five experimenters, representing Britain, France, Germany, and the United States, from 1878 to 1885, tested the competence of the ileocecal valve. "Heschl made a number of experiments on the cadaver and satisfied himself that the ileocecal valve serves as a safe and perfect barrier against the entrance of fluids from below," wrote the author of one review. W. W. Dawson of the Medical College of Ohio put the ileocecal through its paces on thirteen cadavers; in twelve, the valve held strong. The transcript of the thirteenth cadaver demonstration is printed in an 1885 issue of the *Cincinnati Lancet and Clinic.* ("From your seats, . . . you see the colon expanding as the fluid enters.") This one, he concludes, was an anomaly. "The valve was doubtless imperfect." But the showmanship flawless.

It seems fair to say that it takes an unnatural volume of liquid, under unnatural pressure, to breach the heroic ileocecal valve and enter the small intestine from the rear. It takes, perhaps, a Joy-Beauty-Life colonic irrigator. In their fervor to rid the body of fecal residues, devotees of internal bathing were flushing the dread residues higher up into the gut, away from the colon—a region of the anatomy that does relatively little absorbing—and right on into the one that evolved specifically for the job, the small intestine.

You may be wondering why the minds of medicine so assiduously concerned themselves with the matter. Were they drawn to it simply as lecture hall spectacle? Not entirely. The experiments aimed to resolve a lingering medical debate over the value of "feeding per rectum."

# 15

## Eating Backward

### IS THE DIGESTIVE TRACT
### A TWO-WAY STREET?

$A$s FAR BACK as ancient Egypt and as recently as 1926, patients unable to keep their food down would be given their food up. The "nutrient enema" was a last resort for people who, the thinking went, would otherwise starve. As unlikely as it may sound, the practice was broadly accepted in the medical community, so much so that ready-made preparations were available for purchase. You would see them advertised in the pages of journals, complete with the occasional customer testimonial (as from the satisfied 1859 patient for whom rectal coffee* and cream

---

\* But not boiling hot coffee. The contemporary fad for coffee enemas has sent more than one person to the emergency room with a partially cooked colon. I first heard about this from a veteran ER nurse. "You have no idea what people will do to themselves," she wrote in an e-mail. "Forget to remove the potato that you used as a pessary until you noticed a vine sprouting between your legs? Decided to do your own nose job at the bathroom mirror and replace the cartilage with a leftover piece from last night's chicken dinner? You have no idea."

"relieved the sense of 'famishing thirst' better than any other injection").

President James Garfield was the poster boy of rectal feeding. In 1881, Garflield's liver was pierced by an assassin's bullet and shortly thereafter inoculated with a dose of bacteria from the unwashed fingers and instruments of Dr. D.* W. Bliss. From August 14 to the time of Garfield's death on September 19, the dwindling, retching head of state, on Bliss's orders, was fed nothing but nutrient enemas prepared in the dispensary of the United States surgeon general.

Here is the recipe for Assistant U.S. Surgeon General C. H. Crane's Rectal Beef Extract: "Infuse a third of a pound of fresh beef, finely minced, in 14 ounces of cold soft water, to which a few drops of muriatic acid and a little salt . . . have been added. After digesting for an hour to an hour and a quarter, strain it through a sieve." The yolk of an egg was then added, along with 2 drams of Beef Peptonoids and 5 drams of whiskey.

The nice thing about cooking for someone who can't taste the food is that the same dish can be served over and over without complaint. Or without the usual complaint. A downside to eating rectally is that body heat quickly leads to rot and reek. Presi-

---

* The *D* stood for "Doctor." Garfield's doctor was Dr. Doctor Willard Bliss. For reasons lost to time, Bliss's parents named their boy after a New England physician, Dr. Samuel Willard. It would seem they mistook the doctor's title for his first name, for rather than naming their son Samuel Willard Bliss, as the custom would dictate, they christened him Doctor Willard Bliss. Perhaps to simplify his life, the boy went into medicine—despite a seeming shortage of aptitude and professional ethics. In addition to allegedly hastening Garfield's death (and then submitting a bill for $25,000—around half a million in today's currency), Bliss is said to have employed untrained cabinet members' wives as nurses. Conveniently, no matter what happened, even were he stripped of his medical license, he would always be Doctor Bliss.

dent Garfield and his nurses endured five days of sulfurous flatus so "annoying and offensive" that egg yolks were stricken from the recipe. Beef blood was likewise to be avoided; one physician lamented that the odor produced by decomposing blood was "so offensive as to pervade the whole house." Bouillon, another common rectal menu item, also created optimal conditions for bacteria. (Before agar was widely used for laboratory cultures, a medium of choice was beef broth.) The enema-fed rectum was a highly efficient incubator, an in-house petri dish.

What's worse, proceeding too quickly could trigger the more traditional goal of the enema. (I suppose it wasn't that far removed from feeding a baby. Though where do you hang the bib?) "I need hardly say," wrote a learned contributor to the *British Medical Journal* in 1882, "that the rectum should be empty when a nutrient injection is to be given." A before-dinner enema of the cleansing variety was recommended.

As a way around the problem, food could be mixed with wax and starch to form a suppository. An additional advantage of this, wrote Bliss in *Feeding per Rectum*,* was that patients could manage their own feeding and need not be confined to the hospital. "The convenience of this method is very great," he enthused. It was the Clif Bar of rectal alimentation. Bliss followed with a caveat: "In some cases, owing to irritability of the rectum, the whole suppository has been returned." In the history of medicine, has a gentler euphemism ever been coined for the act of excretion? *Excuse me, here you go, I'm returning this?*

Eventually Heschl and Dawson and the others came along, hosing their cadavers and publishing their papers. The ileocecal

---

* Why an entire book about rectal alimentation? Because, said Bliss, it is "more interesting than any romance."

valve experiments made it clear: the small bowel—the homeland of nutrient absorption—was, under normal, nonhydraulic circumstances, unreachable via reverse passage. This is why the meat preparations tended to include some minced pancreas. The hope was that the pancreatic enzymes would break down proteins into something more readily absorbed by the colon and rectum.

Did rectal feeding provide nourishment or just hydration? What—and how much—*was* being absorbed? A round of experiments got under way, and it soon became clear that the colon and rectum were incapable of absorbing large molecules: fats, albumins, proteins, all of it was returned a few days later. Salt and glucose, some short-chain fatty acids, a few vitamins and minerals, these things were retained to a certain extent. And little else. Ninety percent of nutrient absorption takes place in the small intestine. Rectal meals could postpone death, but it was an exaggeration to say they sustain life.

Interestingly, the Vatican proposed a similar experiment in the 1600s. The Church sought an answer to the nagging question "Does rectal consumption of beef broth break one's Lenten fast?" This was a subject of some controversy within the Church. Pharmacists of the day were turning a brisk business administering bouillon enemas to nuns and other pious, peckish Catholics who found that this helped them make it to lunch. The Vatican rules on fasting define food as "something digestible, received from outside into the mouth and passed by swallowing into the stomach." By this definition, an enema does not technically break one's fast.*

---

* The priestly handbook *The Celebration of Mass* helpfully enumerates other substances that may enter the digestive tract without technically breaking one's fast: gargled mouthwash; swallowed pieces of fingernail, hair, and chapped skin from the lips; and "blood that comes from . . . the gums."

Enema madness in the convents was forcing the Vatican to reconsider. An experiment was proposed whereby volunteers would be fed strictly by rectum. If they survived, the enema would have to be considered food and therefore banned. If they didn't, the definition would remain as is, and some vigorous penance would be in order. In the end, nobody volunteered and the nuns continued, wrote Italian medical historian A. Rabino, to "welcome the clysters in their cells with tranquil conscience."

O WING TO THE limited talents of the colon as an organ of absorption, perfectly good nutrients are daily discarded. The small intestine has time to absorb only so much before passing the goods along to the colon. Bacteria in the colon break down what they can, creating vitamins and other nutrients in the process, but because the colon isn't as well set up to absorb the locally produced bounty, some of it is excreted.

This topic came up during a conversation with pet-food scientist Pat Moeller, of AFB International (and chapter 2). Moeller had offered an explanation for the disconcerting canine habit of autocoprophagia. "If you think about it"—and, improbably, we were—"a dog that eats its stool, in some cases, may be getting missing nutrients" by running a meal through the small intestine twice.

In some neighborhoods of the animal kingdom, your own is a regular second course. For rodents and rabbits, in whom vitamins B and K are produced exclusively in the colon (by bacteria that live there), the self-manufactured pellet is a large, soft daily vitamin. Which brings us to Richard Henry Barnes and a little-known chapter of nutrition history.

Richard Henry Barnes was the dean of the Graduate School

of Nutrition at Cornell University from 1956 to 1973, the president of the American Institute of Nutrition, and the first academic to formally address the consumption of shit. I found a photograph of Barnes taken around the time his "Nutritional Implications of Coprophagia" ran in *Nutrition Reviews*. His blond hair had receded from his temples and was combed flat against his skull. His glasses were the two-toned horn rims popular in the late 1950s. Ed Harris could play the part. Barnes did not appear to be in any part an iconoclast. "One of the qualities I respected most in Dick," a colleague reminisced in a Barnes obituary, "was his complete open-mindedness and objectivity in dealing with . . . socially and politically sensitive questions."

Barnes's original interest in rodent autocoprophagia grew out of efforts to prevent it. Like other nutritionists of his day, Barnes was frustrated to find his carefully controlled diet studies repeatedly undone by his subjects' menu substitutions. Experimenters before him had tried building cages with wire-mesh flooring that allowed fecal pellets to drop through. This proved to be of limited use because, quoting Barnes, "feces are consumed as they extrude from the anus." Rats on mesh floors still managed to consume anywhere from 50 to 65 percent of their "total output."

Presently Barnes became more interested in the inputting of output than in the elements of nutrition he'd originally set out to study. "The contributions of coprophagy in rats as a means of making available the nutrients that are synthesized in the lower intestine has remained one of the major nutritional mysteries of our time," he wrote in a 1957 paper funded by, holy shit, the National Science Foundation (NSF).

Barnes began by documenting the precise extent to which egesta made up his rats' daily fare. This he did by fashioning "feces collection cups" from the necks of small plastic bottles and fitting them over the rat's tail and rear end. And here we get a glimpse of the industriousness and creativity of Richard Henry Barnes. Part of that NSF grant went to cover the cost of a band saw, Forstner drill bit, wood chisel, Scotch tape, metal bands, rubber tubing, and three different sizes of plastic bottles from the Wheaton Plastics Company. Daily collections were emptied from the cup and served to the animal in its feed jar, which I like to picture with a silver warming cover, lifted with a flourish by Barnes himself. The rats, Barnes found, ate 45 to 100 percent of what they'd excreted each day. If you prevent a rat from doing this, Barnes further noted, it will quickly become deficient in vitamins B5, B7, B12, and K, thiamine, riboflavin, and certain essential fatty acids.

Four years later, B. K. Armstrong and A. Softly, scientists with the Department of Biochemistry and the Animal House at Royal Perth Hospital, showed that preventing rats from eating their first round of excreta severely stunted their growth. Over the course of a forty-day experiment, young rats thusly stymied gained just 20 percent of their starting body weight, while an unhindered control group gained 75 percent. (Both groups ate all their other food as well.) Armstrong and Softly developed their own unique method of restraint, eschewing the Barnes technique. "To eliminate the necessity for continual emptying and replacement of fecal cups, we have used a jacket to prevent the rat from reaching its anus."

"Used a jacket" is a humble understatement. A pattern (included in the journal paper) was drawn up and soft purse leather purchased. "A V-shaped tail cleft was trimmed to clear the penis or vagina. The laces were adjusted to give a firm, but not tight fit, and the string was tied at the tail in a knotted bow. Final adjustments were made with fine scissors." It all sounds very Stuart Little until you turn the page and come upon plate 1: "Rats wearing jackets to prevent coprophagy." The leather is black, and the jacket, actually a vest, is laced along the animal's midline like a corset. An attached black leather collar completes the look. Suddenly "restraint" took on a whole new flavor, and you began to wonder what went on after hours at the Animal House.

Barnes likened autocoprophagia to rumination: another strategy to get the most out of one's meal. Cows will rechew and reswallow the same mouthful forty to sixty times, greatly increasing the surface area that rumen bacteria have to work with and extracting maximum nutritive value. In fact, one of the alternate terms for autocoprophagia is "pseudo-rumination." No doubt the word was coined by a rabbit fancier. Rabbits are diehard autocoprophagics, and their owners seem a little uncomfortable with it. In rabbit circles, the first round's larger, softer fecal pellets* have a special, non-fecal-sounding name: cecotropes. "Cecotrophy, not Coprophagy," tuts a heading in one journal paper.

"It seems likely that most nonruminant species have a voracious appetite for feces," Barnes bravely continued. "This practice is so normal to their nutritional behavior that the . . . large intestine should rightfully be considered as functionally positioned ahead of the absorptive region of the intestinal tract." In other words, a second visit to the small intestine is the true end point for absorption.

I will buy that autocoprophagia is, as Barnes put it, "a normal practice for . . . rats, mice, rabbits, guinea pigs, dogs, swine, poul-

---

* Given the situation with rabbits and their fecal pellets, you would think the producers of commercial rabbit food would have steered clear of the word *pellets*. When, say, the Kaytee brand boasts, "Quality, nutritious ingredients in a pellet diet that rabbits love," I don't necessarily picture a bag of kibble.

try, and undoubtedly many others." But Richard: "Most nonruminant species"?

Let's check in first with our closest cousins. I e-mailed Jill Pruetz, the Iowa State University primatologist whose work with chimpanzees in the Fongoli River region of Senegal I profiled for a magazine in 2007. By coincidence, Pruetz and her colleague Paco Bertolani had just submitted a paper on the topic. "I don't like to think of the Fongoli chimps as shit-eaters," she wrote back, "but what are you going to do?" For one thing, you call it "seed reingestion." Technically speaking, this is accurate. Fongoli chimps don't, as they say, "consume the dung matrix." They "excrete a faecal bolus into one hand and then extract the seeds from it with the other hand or with the lips." You may be pleased to note that when they are done they "clean their lips by rubbing them on the bark of trees."

Pruetz's team observed seed reingestion only during the span of weeks when baobab and Fabaceae seeds are too hard to chew. During this time, it takes a second run through the digestive tract to dissolve the hulls and release the proteins and fats in the kernel. Women in the Tanzanian Hadza tribe use a similar technique, harvesting softened baobab seeds from baboon dung, washing and drying them, and pounding them into a kind of flour.

Before you get all high and mighty on the chimps and the Hadza, you should know that the most expensive coffee beans in the world—at upwards of two hundred dollars a pound—are those that have passed through the digestive tract of the civet, a catlike animal native to Indonesia. The animal's digestive enzymes are said to alter the taste of the beans in a pleasing manner. The trade is lucrative enough to have spawned a market for counterfeit civet dung, crafted from ordinary undigested coffee beans, a dung matrix of similar consistency, and glue.

Though seed reingestion is most prevalent on the savannah, where food is scarcer, it also happens in the rain forest. Pruetz's paper cites the work of a team of researchers who observed coprophagy in wild mountain gorillas. At a loss to explain the behavior, given the relative bounty of the surroundings, the researchers suggested that it might have been done for the same reason people reach for the Cream of Wheat on a midwinter morning. "They proposed," Pruetz wrote to me in an e-mail, "that mountain gorillas might like to eat something warm during periods of cold temperatures or heavy rain."

And now, with all apology, it's time to move on to *Homo sapiens*. A 1993 study of "humans behaving in a manner similar to nutrient-deficient animals" involved three institutionalized patients, Bart, Adam and Cora, all with profound developmental disabilities. Charles Bugle and H. B. Rubin successfully broke the trio's autocoprophagia habits by feeding them a nutritional supplement drink called Vivonex. The authors speculated that this population "often has multiple handicaps and something may be missing that makes it more difficult to digest or metabolize all the nutrients in the diet they are served." Whether or not this is true, a glass of Vivonex is preferable to some of the alternative strategies tried by staff at other institutions. In particular, that of the team who "treated . . . coprophagia and feces-smearing by making a shower contingent upon the absence of feces." You can see where that could go south pretty fast.

THERE IS ONE class of substances that the rectum, even today, is occasionally called on to absorb. Drugs take effect faster this way than by mouth, partly because they bypass the stomach and liver. Opium, alcohol, tobacco, peyote, fermented agave sap, you name it—it's been taken rectally. In the case of certain South American hallucinogens, rectal indulgence also allows one to sidestep vomiting that accompanies the oral route. Considerably enlivening the pages of *Natural History* in March 1977, Peter Furst and Michael Coe described the heretofore unrecognized prominence of the "intoxicating enema" in classic Mayan culture. The discovery came about with the examination of a painted Mayan vase from circa 3 A.D. that had previously been hidden away in a private collection. The decorative embellishments feature a man in an elaborate pointy hat but no pants, crouched like a cat, hind quarters raised, while a kneeling consort holds a tubular object to his anus. Another man squats, administering to himself.

Access to the vase brought a thunderclap of realization. "Previously enigmatic scenes and objects in classic Maya art" suddenly made sense. Furst and Coe give the example of a small clay figurine, found in a tomb, of a squatting man reaching back as though to wipe himself. Experts had been puzzled. Why would family members bury a loved one with the Maya equivalent of Manneken Pis? Now it was clear. The man was on a ritual bender. Images on the vase no doubt also helped crack the enigma of what had appeared to be rustic, hand-hewn turkey basters—hollow bones with animal or fish bladders attached at one end—turning up at archaeological digs all over South and Central America. "South American Indians," observe Furst and Coe, "were the first people known to use native rubber-tree sap for bulbed enema syringes."

Is it not possible that the images on the vase depict a simple laxative procedure? Furst and Coe address this, insisting that only partakers of the "Old World enema" were concerned with constipation. (Sometimes to excess. The authors note that Louis XIV had more than two thousand clysters during his reign, sometimes "receiving court functionaries and foreign dignitaries during the procedure." The Louis passion for the syringe can be traced through the lineage as far back as XI, who had enemas administered to his dogs.)

The southern route has advantages as well for administering poisons. Bypassing the taste buds—and the court taster, if such an entity actually existed—allowed murderers to get away with a higher dose. Some historians believe the Roman emperor Claudius was killed in this manner, at the behest of his fourth wife, the fetching and far younger Agrippina. Ostensibly the motive was political. Agrippina was in a rush to install her son from a previous marriage as Rome's emperor. There was also this, courtesy of Suetonius: "His laughter was unseemly and his anger still more disgusting, for he would foam at the mouth and trickle at the nose; he stammered besides and his head was very shaky." And this, from the September 5, 1942, issue of the *Journal of the American Medical Association*: "The emperor Claudius . . . suffered from flatulence."*

By far the oddest reverse delivery on record is the holy-water enema. The first reference I came upon, a passing mention in an art journal, suggested that the holy-water clyster was a routine weapon in the exorcist's arsenal. This made a certain amount of sense: Why sprinkle the possessed with holy water when you can pump it right up inside them? Seeking to verify the practice, I e-mailed the public relations office of the United States Confer-

---

* Which explains the otherwise curious legislative decision to pass an edict that "no Roman need feel reticent about passing flatus in public."

ence of Catholic Bishops, the stateside headquarters of the Catholic Church. Naturally this went unheeded. Returning to the art journal, I consulted the article's references, ordered a copy of the cited paper, and hired a translator, as it had been published in an Italian medical journal.

The holy-water enema, by this account, was an isolated case, involving Jeanne des Anges, the mother superior of an Ursuline convent in Loudun, France, in the early 1600s. Des Anges claimed that the parish priest, a raffish, high-ranking charmer named Urbain Grandier, was appearing to her in her dreams, caressing her and attempting to seduce her. He seemed to be having some measure of success, as the contemplative quiet of the convent was being shattered by the mother superior's nightly shrieks of sexual frenzy. An exorcism was promptly ordered.

Why would one administer the blessed liquid rectally instead of simply having the possessed drink a glass of it? One explanation is that the original Roman Catholic rite for the Blessing of the Holy Water included adding salt to the water. Regardless of the origins of the practice, this had the effect of rendering it undrinkable.*

---

* Is drinking holy water allowed? Clear-cut answers are elusive. One priest I contacted pointed out that holy water is baptismal water, meant for blessing and dunking, not drinking. Another, however, directed me to the website of McKay Church Goods, which sells five different models of "Holy Water tanks." These are six-gallon freestanding dispensers with push-button spigots, along the lines of the office water cooler but with a cross on top. There are definitely parishioners who drink it, and priests who wish they wouldn't. St. Mary's Parish in Cutler, California, has had both. In 1995, Father Anthony Sancho-Boyles, to discourage tippling, resorted to the old practice of adding salt to the holy water. The following Sunday a woman complained, saying that she used the holy water to make coffee in the mornings, and now her coffee tasted funny.

Here's the other reason: "After many days in which the priest tried to dispel the devil, he learned from the possessed mother superior that the devil had barricaded himself inside . . ." Here my translator stopped. She leaned closer to the photocopied pages and traced the words with her finger. ". . . *il posteriore della superiora*. Inside her butt!"

Sensing that the situation had progressed beyond his expertise or comfort level, the exorcist called for outside help in the form of a pharmacist, "Signor Adam," and his traveling syringe. (Enemas in those days were the purview of pharmacists and comprised a sizable percentage of their income.) Mr. Adam "filled up the syringe with holy water and gave the miracle clyster to the mother superior, with his usual skill." Two minutes later the devil had vamoosed.

Books about the Loudun fracas, including a 1634 translation of an account by "an eyewitness," include no mention of Mr. Adam or rectal exorcism, but they do serve to flesh out the story. Grandier was convicted of sorcery and burned at the stake, and most sources agree he'd been framed by des Anges, acting in cahoots with a rival priest. The "possessions" continued for several years after the execution, spreading to sixteen other nuns and turning the convent into a local tourist attraction, and understandably so: "They . . . made use of expressions so indecent as to shame the most debauched of men, while their acts, both in exposing themselves and inviting lewd behavior . . . would have astonished the inmates of the lowest brothels in the country."

In the words of my translator Rafaella, responding to the material I had engaged her to read, "I am sorry, but nuns should be allowed to have sex." Or at least an occasional holy-water enema.

• • •

$A$ROUND THE TIME doctors took to serving dinner through "the other mouth"—as Mütter Museum curator Anna Dhody has called the anus—a phenomenon called antiperistalsis began cropping up in medical journals. This was distinct from the fleeting reverse-peristaltic lurch of vomiting, wherein the small intestine squeezes its contents backward into the stomach, whose sphincters have opened to grant through-passage. That is normal.

This is not. "For eight days this person, at least once and sometimes twice in twenty-four hours, vomited veritable feces, solid, cylindrical, of a brown color and with the normal faecal odor, coming evidently from the large intestine." The patient was a young woman, admitted to a hospital in Lariboisière in 1867, under the care of a Dr. Jaccoud, for a bout of hysterical convulsions. This was not the first alleged case of "defecation by the mouth." Writing in 1900, Gustav Langmann summarized eighteen case reports of widely varying plausibility.

Jaccoud assumed his patient had an intestinal obstruction. When digesta backs up to the point that it threatens to burst the pipes, an emergency measure called "faeculent vomiting" kicks in. But the material in that case is highly liquid, coming, as it does, from the small intestine. A well-formed stool does not exit the upper end of the colon.

Besides, the woman showed no symptoms of a life-threatening obstruction. "Apart from the passing disgust which followed the act," Jaccoud noted, "the patient ate as usual and continued in her ordinary health." Things simply appeared to be running in reverse. Jaccoud's colleagues suspected he'd been had. Defecation by mouth was a showstopper in the tradition of stomach snakes or the birthing of live rabbits (which turned out to have been sequestered in the woman's skirts). Experts would travel great distances to observe a spectacle of this caliber. For the lonely or neglected patient who craves attention, it was just what the doctor ordered.

In 1889, Gustav Langmann put an alleged reverse-defecator to the test. A twenty-one-year-old schoolteacher, identified as N.G., had been admitted to the German Hospital of New York on and off for over a year, with the complaint of repeated spells of vomiting. On May 18 of that year, witnesses reported she threw up "hard scybala" the size of malted-milk balls. "It seemed," wrote Langmann in his paper, "to be a favorable time to experiment in regard to the carriage of substances from the rectum to the mouth."

At 11:01 A.M., Dr. Langmann injected just under a cup of water tinged with indigo dye into the woman's rectum. "Blue feces took its natural course," which is to say it emerged from the customary direction. A few days later, a nurse reported having discovered "some hard feces, wrapped in paper," under the woman's pillow. Langmann reports that she later tried her "tricks" at two other medical facilities.

Human beings do not defecate through the same orifice they eat with. That is a feat reserved for the cnidarians*—sea anemones and jellyfish being the best-known examples.

---

* Pronounced "nidarians." But not to be confused with the Nidarians, elite players of the online game Remnants of Skystone. The cnidarians are covered with stinging cells. The Nidarians are covered with purple mold and are entitled to "two extra attacks per class," "a 10 percent discount when using Spores," and "more baking and brewing possibilities."

Contributing to the confusion about "antiperistalsis" was the fact that the normal waves of intestinal peristalsis run in both directions. It's a mixing function. The better the digesta circulate, the more nutrients come in contact with the villi. Though the net movement is forward, it is, as Mike Jones put it, a "two-steps-forward-one-step-back phenomenon."

Look up *antiperistalsis* in the medical literature, and you will come across a brief, curious phase in the history of surgery. In 1964, a team of northern California surgeons took an ambitious and iconoclastic approach to curing chronic diarrhea and improving absorption. To slow forward transit through the small intestine, they removed a six-inch segment of it, turned it around, and stitched it back in place.

Jones points out that the body has a tendency to rewire itself as it sees fit. A 1984 study followed four patients who'd had the operation. Within two years, the diarrhea had returned.

For milder cases, a shift of perspective may be helpful. "When I see a patient with a little bit of diarrhea," Michael Levitt told me, "I say, 'Just be happy you're not constipated.'"

# 16

## I'm All Stopped Up

### ELVIS PRESLEY'S MEGACOLON, AND OTHER RUMINATIONS ON DEATH BY CONSTIPATION

*L*ENIN'S TOMB IS unusual among public memorials in that it displays the man's actual remains. As such, it attracts not only those who wish to pay respect, but others, like me, who are simply curious. Either way, death demands a respectful silence, and one cannot easily distinguish mourner from gawker. I was reminded of Lenin's tomb when I visited the Mütter Museum, in Philadelphia, to view the remains of a man identified as J.W. There was the glass case and the careful curatorial lighting, the transfixed but largely unreadable faces of the visitors, the general hush and horror of it.

The J.W. vitrine doesn't exhibit a corpse—just a colon. That this glass case is not much larger than the one that holds Lenin tells you two things: Vladimir Ilyich Lenin was a small man, and the colon of J.W. was enormous: twenty-eight inches around at its most distended point. I remember standing there thinking, *It wears the same size jeans as me.* A normal colon, perhaps three inches around, has been laid alongside for scale.

What happened here? Hirschsprung's disease. As J.W.'s embryonic self was laying down nerves along the length of the colon, the process petered out. The final stretch was left without. As a result, peristalsis—the wave of contraction and dilation that moves things through the gut—stops right there. Digesta pile up until the pressure builds to a point where it shoves things through. The shove might happen every few days, or it might take weeks. Just behind the dead zone, the colon becomes overstretched and damaged—a floppy, passive, swollen thing. The megacolon may eventually take up so much room that it begins to bully other organs. Taking a deep breath is a struggle. J.W.'s heart and lungs were thrust upward and outward to the point where they pushed the ribs aside and began jutting horizontally from the torso.

Without surgery, a megacolon like J.W.'s will prevail. If the specimen is spectacular enough, it finds its way to a museum, earning a toehold in medical history while the man himself fades to obscurity. This was the case, too, with the megacolon of a Mr. K., written up in the *Journal of the American Medical Association* in 1902. In a photograph that accompanies the article, the organ lies on what appears to be a hospital bed, as though it grew so big that it eventually eclipsed Mr. K. entirely and the doctors and nurses took to caring for it in his place, changing the sheets, bringing meals on trays, putting bendy straws in its ginger ale. All we know about poor Mr. K. is that he lived in Groton, South Dakota. Everything else has been subsumed by the details of the autopsy and a frightful chronology of doctor-assisted evacuations. From a medical aside, we glean that Mr. K. had a family and that they seemed to care about him: "June 22, the report was received that he had passed an ordinary pailful of feces. . . . There was much rejoicing in the family."

Anna Dhody, the Mütter Museum curator, led me down to the basement* to see what we could learn about J.W. the man. The file holds a reprint of a paper presented at the College of Physicians of Philadelphia on April 6, 1892, by Henry Formad, Demonstrator of Morbid Anatomy. On top of overseeing the "rather voluminous autopsy," Formad had interviewed J.W.'s mother. The woman recalled that "disturbances in defecation" and abdominal swelling had been evident by age two, suggesting Hirschsprung's. J.W. began working at sixteen, first in a foundry and later a refinery. All the while, his belly continued to swell. In a photograph taken shortly before he died, he stands in a doctor's wood-floor examining room, naked except for hospital slippers, baggy white socks, and a few days' growth of beard. He looks directly at the camera with a demeanor of calm defiance. Imagine the biggest potbelly, the longest overdue triplets, on a meager frame of knobby limbs. The bastard offspring of Humpty Dumpty and Olive Oyl. To better capture the great torso on film, the photographer had instructed J.W. to raise one hand to his head. The cheesecake pose invites you to stare, but everything else says, Look away.

By the age of twenty, J.W.'s physique had grown so peculiar that he was hired by a freak show in Philadelphia's old Ninth and

---

* This was less exciting than it sounds because Dhody keeps the "creepy-tastic" stuff out on display. For example, the necklace of dried hemorrhoids, and the jar of skin (dropped off by the roommate of a compulsive picker, in a Trader Joe's strawberry preserves jar with a note attached: "Please recycle," presumably referring to the jar).

Arch Museum. The museum's first floor housed carnival-style tests of strength and fun-house mirrors, and I imagined J.W. hanging around those mirrors on his breaks, positioning his girth just so and taking in the bittersweet sight of himself as a normally proportioned man. J.W. was exhibited under the carnival name Balloon Man, along with the Minnesota Woolly Baby* and an assortment of other human and animal oddities.

Formad made no reference to J.W.'s emotional state other than to note that he was unmarried and, justifiably, given to drink.

You don't need a megacolon to fall victim to "defecation-associated sudden death," but it helps. At the age of twenty-nine, J.W. was found dead on the floor of the bathroom at the club where he regularly took his dinners. The autopsy report described the death as instantaneous, but there was no evidence of a heart attack or a stroke. Likewise, our Mr. K. died at 2 A.M. while straining, as they say, at stool.

"That's what killed Elvis," said Adrianne Noe. Noe is the director of the National Museum of Health and Medicine, which has its own megacolon, from an unknown party. As we were about to get off the phone, Elvis Presley dropped into the conversation. Noe related that she'd been standing by the megacolon exhibit one day and a visitor told her that Elvis had had one too. The man added that Presley had struggled with constipation his whole life and that as a child his mother Gladys had had to "man-

---

* Oddly, the exhibit chosen for billboarding on the building's exterior was "Young Women Basketball Players."

ually disimpact" him. "He said that's why Elvis was so close to his mother."

A quiet moment followed. "Really."

"That's what he said."

I had heard that Presley died on the toilet, but I'd assumed the location was happenstance, as it was with Judy Garland and Lenny Bruce: an embarrassing setting for a standard celebrity overdose. But the straining-at-stool theory made some sense. With all three autopsies—that of J.W., Mr. K., and E., as Presley's intimates called him—the collapse was abrupt and the autopsy revealed no obvious cause of death. (Though Presley had traces of several prescription drugs in his blood, none was present at a lethal level.) What Elvis's autopsy did unambiguously reveal was a colon two to three times normal size.

At the time it happened, no one pinned Presley's death on his colon or efforts to empty it. It wasn't until years later that Dan Warlick, the coroner on the case, came forward with the megacolon/straining-at-stool theory. Presley's longtime doctor, George "Nick" Nichopoulos, was an eager adopter of Warlick's theory. Nichopoulos had been vilified for overprescribing prescription drugs, and many fans blamed him for Presley's death. He wrote a memoir and made himself available to talk to the press. Few seemed inclined to listen. The reference I came across was on a website hawking herbal constipation remedies. A short piece headlined "Elvis Died of Constipation" had run as the site's lead story (and its middle and last story) under the category Constipation News.

Why didn't the colonic inertia theory come up earlier? Nichopoulos says that at the time, he had never heard of it. Nor had the gastroenterologist who treated Presley in the 1970s. "Nobody knew about it back then," Nichopoulos says.

I recall reading in one of Charles Tyrrell's books that advances in medical knowledge about the colon had, historically, been hobbled by the organ's repulsiveness. Eighteenth- and nineteenth-century dissectors and anatomy instructors would, he claimed, promptly cut the lower bowel out of the cadaver and throw it away, "on account of its scent-bag propensities and nastiness." Michael Sappol, a historian with the National Library of Medicine who has written extensively on the history of anatomy, said he'd heard this too. Leading me to wonder: Does distaste slow progress in treating diseases of the bowel? Does the excretion taboo discourage research, discussion, media attention?

I recall riding a bus in San Francisco, years ago, and seeing a public service ad about anal cancer, "the cancer no one talks about." I had never heard of it, and in the decade and a half since then, I haven't come upon another reference. Until I looked it up, while writing this paragraph, I didn't realize Farrah Fawcett had died of anal cancer. There were references to her ailment as cancer "below the colon." It was like my mother, when I was a kid, calling the vagina "your bottom in front." Up through 2010, anal cancer had no nonprofit society, no one to organize fund-raisers and outreach, no colored awareness ribbon. (Even *appendix* cancer has a ribbon.)* Like cervical cancer, anal cancer is caused by the human papillomavirus; people get it via sex with an infected

---

* It's amber. Because there are more cancers than basic colors, awareness ribbons are like paint chips now: Stomach cancer is periwinkle, ovarian is teal. Colon and rectal cancer are plain blue. They used to be brown (just as the color for bladder cancer awareness is yellow), but some patients objected. A mistake, I say. They could have had brown all to themselves; blue they have to share with Epstein-Barr, osteogenesis imperfecta, victims of hurricane Katrina, drunk driving, acute respiratory distress syndrome, child abuse, baldness, and secondhand smoke.

person, and that seems like something they ought to know when making decisions about using a condom.

Colonic inertia has an even lower profile than anal cancer. And I doubt you'll be seeing bus posters about defecation-associated sudden death any time soon. I imagine the stigma discourages open talk among doctors and patients and people at risk. As Nichopoulos wrote in *The King and Dr. Nick*, "Nothing could have been more embarrassing than having people whispering about his bowel difficulties."

But I have questions. At what point does constipation cross over from unpleasant to life-threatening? How hard would you have to be pushing? How exactly does it kill you? Should certain people be taking stool softeners the way others take baby aspirin?

I know one person who doesn't mind talking about it.

GEORGE NICHOPOULOS LIVES in a leafy Memphis neighborhood of widely spaced homes, on a bend in the road where once or twice a year a drunk fails to notice the curve and crashes his car in the yard of the house across the street. Elvis Presley had the house designed and built as a present for Nichopoulos and his family in the 1970s. You can see that in its day it was modish and luxe: the peaked ceiling with exposed rafters, the massive stone fireplace that divides the open floor plan of the downstairs, the backyard swimming pool.

Nichopoulos escorts me to the sofa. He and his wife, Edna, sit in armchairs to my left and right. The furniture is positioned far enough apart that I hand the doctor my tape recorder, for fear it won't otherwise pick up his words. The coffee table is just out of reach, so that each time I pick up or set down my cup, I have to rise partway from my seat. It's as though the family had been at a loss to fill the expanses of a home designed by someone with far more extravagant taste.

Nichopoulos is recovering from hip surgery. Although he's in his eighties and using a mobility scooter to get around, he doesn't appear much diminished. He's tan and spiffed up, having just arrived home from an appearance at an Elvis Week memorial event. His hair is white, but it is not the sparse, scalp-clinging strands of the nursing-home frail. His stands firm and frames his head like an aura.

I open a folder and pass around the pictures I have of J.W. and of K.'s megacolon resting on its bed. No documents were released from Presley's autopsy, but Nichopoulos has a photo of a similarly proportioned megacolon. He opens his laptop and turns it to face me. I get up to set down my coffee and cross the divide. In the photograph, a surgeon in blue scrubs holds a limp, bloody colon above his head in the triumphant, two-handed pose of an athlete with a trophy cup. Nichopoulos says he thought about including the photo in his book, so people would have a sense of what Presley had been dealing with. "But we knew that Priscilla was not going to allow us to put this in there."

"She goes and puts her nose in everything." A dispatch from the distant island nation of Edna.

I ask Nichopoulos to talk about precisely, medically, what caused Presley's death.

"The night he died he was bigger than usual," he begins. Depending on how long it had been since Presley had managed to empty himself, his girth fluctuated between big and stupendous. He sometimes appeared to be gaining or losing twenty pounds from one performance to the next. "He wanted to get rid of his gut that night. He was pushing and pushing. Holding his breath." As the constipated do. The technical term is the Valsalva maneuver. Let's have Antonio Valsalva, writing in 1704, describe it: "If the glottis be closed after a deep inspiration and a strenuous and prolonged expiratory effort be made, such pressure can be extended upon the heart and intrathoracic vessels that the movement and flow of the blood are temporarily arrested." After a momentary spike, the heart rate and blood pressure plunge as the pressure squeezes off the flow of blood. This is followed by what one paper termed the "after-fling"—the body taking emergency measures to get things back up to speed.

The body's response to this wild, Valsalvic seesawing of the vital signs can throw off the electrical rhythm of the heart. The resulting arrhythmia can be fatal. This is especially likely to happen in someone, like Elvis, with a compromised heart. Fatal arrhythmia is the cause of death listed on Presley's autopsy report. "Probably every physician practicing emergency medicine has encountered tragic cases of sudden death in the lavatory," writes B. A. Sikirov in "Cardio-vascular Events at Defecation: Are They Unavoidable?"

In 1950, a group of University of Cincinnati physicians documented the phenomenon—rather recklessly, I thought—by monitoring the heart rate of fifty subjects, half of them with heart disease, and asking them to "take a deep breath, hold it, and strain down vigorously as if endeavoring to have a bowel movement." No one died, but they could have. It happens often enough that stool softeners are administered as a matter of course on coronary-care wards.

Making matters riskier: bed pans! "The notorious frequency of sudden and unexpected deaths of patients while using bed pans in hospitals has been commented upon for many years," wrote the Cincinnati doctors. Notorious enough for a term to be coined: "bed pan death." Lying flat is as counterproductive a posture as squatting is productive. Squatting passively increases the pressure on the rectum. It does the pushing for you. It also, Sikirov discovered in his study "Straining Forces at Bowel Elimination," makes the task easier by straightening out the recto-anal angle, which I read as "angel." The overall result, purrs Sikirov, is "smooth bowel elimination with only minimal straining."

The other mode of defecation-associated sudden death is pulmonary embolism. The surge of blood when the person relaxes can dislodge a clot in a large blood vessel. When the clot reaches the lungs it can get stuck, causing a fatal blockage, or embolism. A 1991 study found that over a three-year period, 25 percent of the deaths from pulmonary embolism at one Colorado hospital were "defecation-associated." This study's authors took issue with Sikirov over squatting, claiming that descending and rising from a squat raises the risk of dislodging clots in the deep veins of the thighs.

Presley was given laxatives and enemas on an almost daily basis. "I carried around three or four boxes of Fleets," Nichopoulos says, referring to the enema brand and recalling his days on tour with Presley. Getting the timing right was, he says, "a difficult balancing act." Presley sometimes did two shows a day, and Nichopoulos had to schedule the administration such that the treatments didn't kick in while the singer was on stage. This was the low point of Presley's career: the bulky jumpsuit and isosceles sideburns era. His colon had expanded so dramatically that it crowded his diaphragm and had begun to compromise his breathing and singing. Beneath the polyester and girth, it was hard to see the man who had performed on the stage of the Ed Sullivan Theater, his moves so loose and frankly sexual that the producers had ordered him filmed from the waist up. Now there was a different reason to do so. "Sometimes right in the middle of the performance, he'd think, 'I'm passing a little gas,' and it wouldn't be gas," Nichopoulos says quietly. "And he'd have to get off stage and change clothes."

People who saw the Graceland master bathroom would remark on its extravagance—a TV set! Telephones! A cushioned seat!—but the décor was in equal part a reflection of how much time was spent there. "He would be thirty minutes, an hour, in there at a time," Nichopoulos says. "He had a lot of books in there." Constipation ran Presley's life. Even his famous motto TCB—"Taking Care of Business"—sounds like a reference to bathroom matters. (The TCB oath touched on self-respect, respect for fellow men, body conditioning, mental conditioning, meditation, and, according to a group tell-all by Elvis's entourage, "freedom from constipation.")

When Nichopoulos's book came out, a colorectal surgeon named Chris Lahr contacted him. Lahr's specialty is the paralytic colon.* He has excised, in part or in whole, more than two hundred of them, and he surmised that Presley had had one too. When I spoke to Lahr by phone he told me Johnny Cash, Kurt Cobain, and Tammy Wynette had also struggled with obstinate constipation, and he was convinced that they too had stretches of paralyzed colon. But these were also people who struggled with obstinate drug addictions. Opiates, whether they're in the form of heroin or prescription painkillers, drastically slow colon motility (as do, by varying degrees, antidepressants and other psychiatric drugs).

To know which is right—whether it was drugs or genetics behind the King's condition—you'd need some information

---

* He wrote a book on the topic, called *Why Can't I Go?*, which features dozens of defecography stills and close-ups of colon surgery graphic enough that the back cover has a warning. Can I Go Now?

about his childhood. Most people with Hirschsprung's—the main cause of megacolon—are diagnosed as infants or young children. As Mike Jones put it, "They come out of the box that way." If there were truth to the story Adrianne Noe had heard, about Presley's mother having to use her finger on him, that would suggest a hereditary condition like Hirschsprung's. I ask Nichopoulos whether he'd heard the business about manual dis-impaction. Edna volunteers that she'd read that in one of the many Elvis biographies.

Nichopoulos says he looked into it himself. "We were trying to figure out if it was there from birth or whether it was something that came on later. But his mother was gone." Gladys Presley died when Elvis was twenty-two. Presley's father wasn't around the house much when Elvis was a child.

"I wanted to talk to Priscilla about it," he says. Presumably Elvis would have discussed his medical issues with his wife. Nich-opoulos shifts his weight. The hip still causes him pain. "She didn't want to discuss it."

It surprises me that Presley's condition didn't dampen his enthusiasm for food. He so appreciated Edna Nichopoulos's Greek hamburgers that he gave her a ring he'd commissioned, with each of the recipe's ingredients represented by a different-color dia-mond. "Green for parsley," says Nichopoulos when I ask about it, "white for the onion, brown is the hamburger, and yella . . ." Some words are born for the Memphis accent. *Yellow* is one.

"Yella is the onion," says Edna.

Nichopoulos considers this. "Wasn't that the white?"

"No, white's the bread."

"Elaine!" Nichopoulos shouts toward the upstairs. "Can you get the hamburger ring!" Elaine Nichopoulos has been living with her parents, helping out since her father broke his hip.

A few minutes pass before Elaine appears on the stairs. She crosses the living room with a crooked gait, the combined aftermath of a car crash and a fall from a ladder. "Sorry, I was in the bathroom," she says. "I'm sure *y'all* can understand"—*y'all* meaning the freaks down in the living room talking about bowel health.

Elaine sits down on her dad's mobility scooter. She shows me where the pins stuck out of her ankle as it healed. Then she pulls down the shoulder of her shirt. I expect more medical hardware, but it's a tattoo. "Do you like monkeys?" I almost say, but then I get it: There's a monkey on her back. Oxycontin, fentanyl, drugs for chronic pain. On top of everything else, she has fibromyalgia.

". . . and bipolar," her dad chimes in.

She makes a face at him. "No, *you* are."

I ask permission to try on the hamburger ring. "Go ahead," Nichopoulos says. "We've got finger cutters." It's a fabulous object. I love the mix of diamonds and hamburger, glamour and trash. I feel like Elizabeth Taylor and Larry Fortensky at the same time.

ELVIS PRESLEY'S COLON is not on display in a glass case, but you can get a good sense of what it looked like by reading the autopsy section of *The Death of Elvis*. "As Florendo cut, he found that this megacolon was jam-packed from the base of the descending colon all the way up and halfway across the transverse colon. . . . The impaction had the consistency of clay and seemed to defy Florendo's efforts with the scissors to cut it out."

Nichopoulos was at the autopsy and remembers the moment. The clayey material, he says, was barium, administered to prep Presley for a set of X-rays—taken four months earlier. "That barium was . . ." He gestures toward the fireplace. "Just like a rock."

He says the impaction obstructed at least 50 to 60 percent of the diameter of Presley's colon.

In the 1600s, the venerable English physician Thomas Sydenham advocated horseback riding as a remedy for an impacted bowel. I mention this to Nichopoulos, noting that Presley had liked riding well enough to have had a stable built at Graceland.

"That's interesting," he says. "It would certainly loosen it up." Elaine turns the scooter and drives away.

Thomas Sydenham was an uncommonly gentle practitioner. Another of his treatments for intestinal obstruction featured mint water and lemon juice, as if all that were needed to make a man right was a refreshing summertime beverage. "I order, too," he continued, "that meanwhile a live kitten be kept continually lying on the naked belly." The kitten was to remain in place for two to three days, whereupon a dram of something unrecognizable but presumably stronger was prescribed. "The kitten is not to be taken off before the patient begins with the pills."

Sydenham did not explain himself. I was left wondering whether this was an early form of animal-assisted therapy and the kitten's role was simply to help the patient relax while nature took its course. Impactions often resolve on their own. Sydenham once treated an overburdened London businessman by sending him to Inverness to visit a specialist who didn't exist. The patient returned from his weeklong journey vexed but rested and cured.

It's also possible, though unlikely, that the kneading of the kitten's paws was viewed as a kind of therapeutic massage. Around the turn of the last century, massage—or medical gymnastics, as it was also then called—was not uncommonly applied to the obstructed bowel. Here is Anders Gustaf Wide, in the *Hand-Book of Medical and Orthopedic Gymnastics*, discussing the technique of "colon-stroking": "One can at least feel the lower part of the larger intestine and often the hard feces in it and even feel, how, in stroking, these are carried forward in the direction they should go."

Or not. In a 1992 University of Munich study, nine sessions of "colonic massage" failed to speed colon transit time in constipated subjects and nonconstipated controls. The subjects' sense of well-being was monitored throughout the three weeks of treatment, and this too failed to improve. It might have gone differently had the masseuses incorporated some techniques from Anders Gustaf Wide—"anal massage," for instance, wherein "small circular strokings are made to each side alternately with tremble-shaking round the anus."

Surgeons, too, advocated the use of the hands to dislodge an impaction, though here it was less of a laying-on than a reaching-in. "I propose this evening to demonstrate upon the cadaver some phases of bowel exploration," began our friend W. W. Dawson, the professor of surgery from the Medical College of Ohio, whom we met in a previous chapter. The year was 1885. Dawson introduced his assistant, Dr. Coffman, to the gathered crowd and then turned to face the examining table. "The subject, you see, is a female." We're going to skip ahead to item 2 on the agenda: "How far can the hand be introduced?" The "patient" was rolled on her back with the thighs raised and the knees bent. The position is known as the lithotomy position, or the missionary position, depending on whether you are taking things out or putting them in. In this case, it was a bit of both. "Dr. Coffman now introduces his hand through the anus and presses gently onward and upward." Here Dawson invited the spectators to watch closely, because it was possible to see the bulge of Coffman's hand moving below the body's surface, like a cartoon mole tunneling under the lawn. "Dr. Coffman is able to move his hand with great freedom. You will recognize at once how it would be possible to dislodge . . . impacted feces."*

For the most part, the historical treatment of obstructed bowels took its cues from the world of plumbing. There were, as there are with bathroom pipes, two main strategies: blast it free with water or air (plunge it), or break it up with something metal (snake it). The June 1874 *Atlanta Medical and Surgical Journal* describes Dr. Robert Battey's "safe and ready" method of dissolving "accumulations of hardened feces" by injecting water, as much as three gallons, up the rectum. "So great was the abdominal tension that the water spouted from the anus when pressure was removed," writes Battey of one memorable case, "in a bold stream" two feet high. Battey's lecture was accompanied by a demonstration. A haphazard perusal of the medical journals of the day seemed to indicate, among surgery and anatomy professors, a keen spirit of one-upsmanship that drove lecture hall demonstrations ever farther in the direction of spectacle.

The digestive tract is an intricate, flexuous pipe not easily snaked. Patients had to more or less swallow the snake. For more than a hundred years, swallowing lead shot or metallic mercury, as much as seven pounds, was thought to be a good way to break up an obstruction. The patient was then rolled or shaken, in

---

* Vigorous debate followed, under the italicized heading "Size of the Hand." A hand more than nine inches around is, declares Dr. Charles Kelsey, "unfit for the purpose." Dawson counters that the size of the pelvis must be taken into consideration. "A broad hipped man or woman would admit a ten inch hand readily," and to fix the limits lower would have the effect of "deterring and embarrassing the practitioner who happens to have a large hand." Or four. Dawson also relates the story of a Dr. Cloquet who, "in quest of a glass tumbler," inserted fourteen fingers into a rectum: six of his own, and four belonging to each of two colleagues. The patient's sphincter, if not his dignity, recovered intact.

hopes that the heavy stuff would work its way through the clog. The problem was that the stomach releases its contents gradually, no matter how swiftly they're swallowed. Rather than pushing through the gut in a cohesive front, the metal shot would journey forth in dribs and drabs, appearing on X-rays like an ingested strand of pearls. Just as well. A physician named Pillore, writing in 1776, describes an autopsy he performed on a patient whose small intestine was so weighed down by the two pounds of mercury that had collected in a lump, that a loop of the organ had stretched and sunk down into the pelvis. The man died a month later. Between the mercury, the unresolved obstruction, and the taffy-pulled gut, it's anyone's guess what ultimately did him in.

For a brief span of years, the plumbers stepped aside and the electricians got to work. Like radioactivity in its day, electricity was new and exciting and presumed to cure whatever ailed a person. Galvanic therapy for obstinate constipation—or "obstipation"—entailed passing a mild electrical current through the abdomen. "Efficacious?" an 1871 *British Medical Journal* contributor is quoted in reply to a dubious colleague. "I could hardly get out of the way in time."

The crudest approach to breaking the dam was simply to toss the patient over a hospital attendant's shoulder.* The intestines do not take a fixed position in the human interior, and simple inver-

---

* In related matters: Is it possible to literally knock the shit out of someone? Depends on the shit and who's knocking it. "I had a high school football coach who was an offensive tackle for the Washington Redskins," says gastroenterologist Mike Jones. "He swore to me that Mean Joe Greene hit him so hard he had to go change his pants." Jones added that his coach had had "a bit of the squirts" at the time, and that it would be tough to hit someone hard enough to "knock a solid turd out of him" and not simultaneously kill him.

sion can, in some cases, bring a measure of relief. A Dr. William Lewitt, of Rush Medical College, in 1864 related the case of a man with a tumor in his abdomen the size of "a child's head at term," which was putting the squeeze on his digestive works. "On visiting the patient, we found him suffering intense agony from pain in the abdomen, with frequent desire to expel flatus from the rectum, which could only be accomplished by standing upon his head and hands, in a perpendicular position." Dr. Lewitt gave his title as Demonstrator of Anatomy, and I imagine it took all the restraint he had not to pack the man up and bring him down to the lecture hall for a demonstration.

The treatment of last resort was surgery. If a blockage could not be shaken, stroked, hosed, or zapped into submission, it was likely to be excised. Surgery in the pre-handwashing, pre-glove-donning era bore a sobering risk of infection. Surgery on the bacteria-laden colon, all the more so. Horrifyingly, colectomy was being performed not just for life-threatening impactions, but as a treatment for constipation and its spurious consequence: autointoxication. What better way to speed digesta through the body than by shortening the chute? Scottish surgeon Sir Arbuthnot Lane, the operation's inventor and vociferous champion, began with "short circuits," removing a span of a couple feet. Soon he moved on to total colectomy, removing basically healthy colons and stitching the end of the small intestine directly to the rectum. If diarrhea can be considered a cure for constipation, he may have done his job, but in the process he put his patients at risk of nutritional deficiencies. As we learned from the coprophagic rodents of chapter 15, the colon—via the metabolic labors of its bacteria—produces not just feculant putridity, but valuable fatty acids and vitamins.

Lane was a raging coprophobe. The normal variances of skin color that you or I would attribute to race or time spent in the sun, Lane perceived as staining from fecally poisoned blood. One patient's "yellowish-brown complexion" disappeared, he noted with pride, a month after her surgery. "She has lost almost all her brown colour," he wrote of another woman. Lane went so far as to deem the colon a useless structure and a "serious defect in our anatomy."

It takes a sizable sum of arrogance and ignorance to second-guess human anatomy and the evolutionary fine-tuning that produced it. The colon that Lane would so cavalierly lop from his patients' interiors is more than a simple waste-storage facility. The bacteria feared and despised by the likes of Lane and Tyrrell and Kellogg—the germs that live and thrive and ply their trade within our waste—are not only harmless, they are critical to good health.

# Bristol Stool Chart

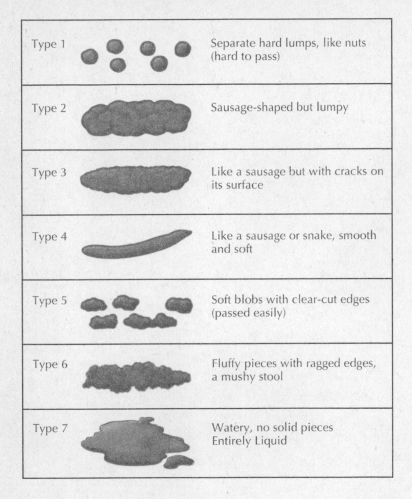

| Type 1 | Separate hard lumps, like nuts (hard to pass) |
| Type 2 | Sausage-shaped but lumpy |
| Type 3 | Like a sausage but with cracks on its surface |
| Type 4 | Like a sausage or snake, smooth and soft |
| Type 5 | Soft blobs with clear-cut edges (passed easily) |
| Type 6 | Fluffy pieces with ragged edges, a mushy stool |
| Type 7 | Watery, no solid pieces Entirely Liquid |

# 17

# *The Ick Factor*

## WE CAN CURE YOU, BUT
## THERE'S JUST ONE THING

*I*T IS A standard party invitation in most respects. There's a street map of the neighborhood, the address and time of the party, and some friendly encouragement to bring the whole family. The decorative elements, though, are unusual: a cutaway illustration of the interior of the human colon, its parts neatly labeled. Above this, in a festive typeface, it says, "Gut Microflora Party!" The host is Alexander Khoruts, a gastroenterologist and associate professor of medicine at the University of Minnesota. Along with the usual complement of colonoscopies and dyspepsia consults, he performs transplants of colon bacteria—aka gut microflora.

Almost everyone gathered at the party this evening is involved with this work. There is Mike Sadowsky, coeditor of the textbook *The Fecal Bacteria* and Khoruts's research partner. Leaning into the buffet is Matt Hamilton, a University of Minnesota postdoc student who prepares the matter for transplant. Matt is spooning Khoruts's homemade Russian red beet salad onto a plate, enough of it that a nurse tells him he's going to "look like a GI bleed" tomorrow.

The nurse admires a platter of chocolate-covered whole bananas, one of the thematically appropriate desserts created by Khoruts's thirteen-year-old. James is very much his father's son, intelligent and cultured, with a sly sense of humor. He plays classical music on the grand piano in the living room and would like to write novels one day. The nurse asks James what number the desserts* would be on the Bristol Stool Scale. He replies without hesitating—4 ("like a sausage or snake, smooth and soft").

It's tough to find an inappropriate mealtime conversation with this group—not because they're crass or ill-mannered, but because they view the universe of the colon very differently than the rest of us do. The interactions between the human body and its gut microbiome—as our hundred trillion intestinal roomers are collectively known—is a hot research area of late. For decades, medical investigators have looked at the role of food and nutrients in disease treatment and prevention. That has begun to seem simplistic. Now the goal is to tease apart the interactions between the body, the food, and the bacteria that break down the food. One example is the cancer-fighter du jour: the polyphenol family, found in coffee, tea, fruits, and vegetables. Some of the most beneficial polyphenols aren't absorbed in the small intestine; we depend on colonic bacteria to metabolize them. Depending on who's living in your gut, you may or may not benefit from what

---

* Not one was eaten. Research by University of Pennsylvania disgust expert Paul Rozin would have predicted a 57 percent consumption rate. In his study, subjects were asked whether they'd be willing to eat "fudge curled to look like dog feces." It is a powerful taboo. Twelve percent refused to even touch it, even though they knew it was fudge.

you eat. Or be harmed. Charred red meat has long been called a carcinogen, but in fact it is only the raw material for making carcinogens. Without the gut bacteria that break it down, the raw goods are harmless. (This applies to drugs too; depending on the makeup of your gut flora, the efficacy of a drug may vary.) The science is new and extremely complex, but the bottom line is simple. Changing people's bacteria is turning out to be a more effective strategy for treatment and prevention of disease than changing their diet.

As a member of a culture that demonizes bacteria in general and the germs of other people in specific, you may find it disturbing to imagine checking into a hospital to be implanted with bacteria from another person's colon. For the patient I'll shortly be meeting, a man invaded by *Clostridium difficile*, it's a welcome event. Infection with chronic *C. diff*—to use the medical nickname—can be an incapacitating and sometimes fatal illness.

"When you're fifty-five years old and you're wearing diapers that you're changing ten times a day," Matt Hamilton says, "you're numb to the ick factor." He lifts some stuffed tomatoes to his plate. Matt has the forceful, unabashed appetite of the big, young male.

"For the patient, there is no ick factor," Khoruts adds. "They've been icked out. It's a chronic disease and they just want to be rid of it."

As regards bacteria in general, a radical shift in thinking is under way. For starters, there are way more of them than you. For every one cell of your body, there are nine (smaller) cells of bacteria. Khoruts takes issue with the them-versus-you mentality. "Bacteria represent a metabolically active organ in our bodies." They *are* you. You are them. "It's a philosophical question. Who owns who?"

People's bacterial demographics are likely to influence their day-to-day behavior. "Certain populations in the gut may want you to eat a certain kind of diet or to store energy differently." (A clinical trial is under way in the Netherlands to see if transplants of "donor feces" from lean volunteers will help subjects lose weight;* thus far results are encouraging but undramatic.) Khoruts gave me a memorable example of how behavior can be covertly manipulated by microorganisms. The parasite *Toxoplasma* infects rats but needs to make its way into a cat's gut to reproduce. The parasite's strategy for achieving this goal is to alter the rat brain such that the rodent is now attracted to cat urine. Rat walks right up to cat, gets killed, eaten. If you saw the events unfold, Khoruts continued, you'd scratch your head and go, What is wrong with that rat? Then he smiled. "Do you think Republicans have different flora?"

What determines your internal cast of characters? For the most part, it's luck of the draw. The bacteria species in your colon today are more or less the same ones you had when you were six months old. About 80 percent of a person's gut microflora transmit from his or her mother during birth. "It's a very stable system," says Khoruts. "You can trace a person's family tree by their flora."

The party is winding down. I go into the kitchen to say good night to James and to Khoruts's mirthful, tolerant girlfriend, Katerina. A blender sits on the counter by the sink, waiting to be washed. "Hey," says James. "You missed the chocolate poop smoothies."

That's okay, because I'll be seeing the real thing.

---

* It's called the FATLOSE trial. FATLOSE stands for "Fecal Administration To LOSE weight," an example of PLEASE—Pretty Lame Excuse for an Acronym, Scientists and Experimenters.

• • •

$L$IKE ANY TRANSPLANT, it begins with a donor. "Anyone's will do," says Khoruts. He has no idea which bacteria he's after—which are the avenging angels that bring *C. diff* under control. Even if he knew, there's no simple way to determine whether those species are present in a donor's contribution. Most species of fecal bacteria are tough to culture in the lab because they're anaerobic, meaning they can't live in the presence of oxygen. (Common strains of *E. coli* and *Staph* bacteria are exceptions. They thrive inside people and out, on doctors and their equipment, and everywhere in between.)

The only thing Khoruts requires of donors is that they be free of digestive maladies and communicable diseases. Family members are not the most desirable donors because their medical questionnaire may not be entirely truthful. "You wouldn't necessarily want to reveal to your loved ones that you've been visiting prostitutes." Khoruts is partial to the donations of a local man who, understandably, wishes to remain anonymous. This man's bacteria have been transplanted into ten patients, curing all of them. "His head is getting bigger," deadpans Khoruts. Most of what Khoruts says is delivered deadpan. "In Russia," he told me, "if you smile a lot, they think something's wrong with you." He has to remind himself to smile when he talks to people. Sometimes it arrives a beat or two late, like the words of a far-flung foreign correspondent reporting live on TV.

"Here he is." A tall man, dressed for a Minneapolis winter, lopes down the hallway carrying a small paper bag.

"Not my best work," the man says, nodding hello to me as he hands Khoruts the bag. With no further chitchat, he turns to leave. He does not seem embarrassed, just pressed for time. He's an unlikely hero, quietly saving lives and restoring health with the product of his morning toilet.

Khoruts slips into an empty exam room and dials Matt Hamilton's number. On the morning of a transplant, Matt will stop by the hospital on his way to the Environmental Microbiology Laboratory, where he works and the material is processed. He's usually here by now, and Khoruts is feeling antsy. Anaerobic bacteria outside the colon have a limited life span. No one knows how many hours they can survive.

Khoruts leaves a message: "Hi, it's Alex. The stuff is ready for pickup." He squints. "I *think* that's his number." It would be a provocative message to receive from a stranger. I picture narcotics officers storming the gastroenterology department, Khoruts struggling to explain.

Khoruts has barely hung up when Matt hustles in, all polar fleece and apology. Matt smiles as naturally as Khoruts doesn't. I imagine it is almost impossible to be peeved at Matt Hamilton.

The lab is ten minutes by car. Because Matt is driving fast and the cooler keeps threatening to slide off the backseat, there's a mild tension in the car. The cooler is a tangible presence, somewhere between groceries and an actual passenger. Soon we're circling, looking for parking. Matt resents the waste of time. "If I had organs, they'd give me a parking pass."

The parking turns out to take longer than the processing. The equipment is simple: an Oster* blender and a set of soil sieves. The blender lid has been rigged with two tubes so that nitrogen can be pumped in and oxygen forced out. Two or three 20-second pulses

---

* "Hi Mary—After reaching out to our Oster product team and reviewing the information you sent me, we have come to the conclusion that we prefer not to comment on this subject matter."

on the liquefy setting typically does the trick, and then it's on to the sieves. For obvious reasons, everything is done under a fume hood. Matt chats as he sieves, occasionally calling out a recognizable element: a chili flake, a piece of peanut.*

A decision is made to do a second run through the blender. If the material doesn't flow freely, it can clog the colonoscope and compromise the microbes' spread through the colon. He turns to face me. "So today we've kind of been confronted with what to do when it's a hard, solid chunk rather than an easier mix." It's like *American Chopper* when Paul Sr. or Vinnie addresses the camera to give a summary of what viewers have been seeing.

Finally the liquid is poured into a container with a very good seal and returned to the cooler. It looks like coffee with low-fat milk. There is almost no smell, the gases having all gone up the fume hood. The three of us, Matt and I and The Cooler, hurry back to the car and retrace our route to the hospital.

The transplant patient has arrived. He waits on a gurney in a room made by curtains. Khoruts is in the hallway in his white coat. Matt hands him the cooler. He fills and caps four vials that will be pumped into the patient through the colonoscope. For now, they are laid on ice in a plastic bowl. Khoruts asks a passing nurse where he can leave the bowl while he waits for an exam room to open up. She glances at it, barely breaking stride. "Just don't bring it in the break room."


Lᴉᴋᴇ ᴘᴇᴏᴘʟᴇ, ʙᴀᴄᴛᴇʀɪᴀ are good or bad not so much by nature as by circumstance. *Staph* bacteria are relatively mellow on

---

* Kung pao chicken, if I had to guess.

the skin, presumably because there are fewer nutrients there. Should they manage to make their way into the bloodstream via, for instance, a surgical incision, it's a different story. Receptors and surface proteins allow bacteria to "sense" nutrients in their environment. As Matt puts it, "They're like: 'This is a good spot, we should go crazy in here.'" Gut microflora party! Bad news for the host. Strains found in hospitals are more likely to be antibiotic-resistant, and hospital patients are often immunocompromised and can't fight back.

Likewise *E. coli*. Most strains cause no symptoms inside the colon. The immune system is accustomed to huge numbers of them in the gut. No cause for alarm. Should the same strain make its way to the urethra and bladder, now it's perceived as an invader. In this case, the immune attack itself creates the symptoms—in the form, say, of inflammation.

Even *C. difficile* is not inherently bad. Thirty to fifty percent of infants are colonized with *C. diff* and suffer no ill effects. Three percent of adults are known to harbor it in their gut without problems. Other bacteria may tell it not to make toxins, or the numbers are too small for the toxins to create noticeable symptoms.

The problems often begin when a colon is wiped clean by antibiotics. Now *C. diff* has a chance to gain a foothold. As careful as hospitals try to be, *C. diff* spores are everywhere. And certain conditions in the colon make it easier for *C. diff* to thrive. Diverticuli are pockets along the colon wall, often created by chronic constipation. Like this: If the muscles of the colon have to push hard to move waste along and there's a weak spot in the wall, the matter will follow the path of least resistance. The weak spot will balloon outward and form a small pocket. *C. diff* spores seed the pockets.

Eighty percent of the time, antibiotics clear up a *C. diff* infection. Twenty percent of the time, it comes back within a week or two. The *C. diff* entrenched in diverticula are tough to annihilate; they're the Al Qaeda of the GI tract, hiding out in inaccessible caves. "Antibiotics are a double-edged sword," says Khoruts. "They suppress *C. diff*, but they also kill the bacteria that keep it under control." Every time the patient has a relapse, the chance of another relapse doubles. Infections with *C. diff* kill around sixteen thousand Americans a year.

Today's patient has diverticula that became abscessed. Multiple severe bouts of colitis have caused diarrhea so severe he has had, at times, to be fed intravenously. You wouldn't guess any of this to look at him now, in the exam room. He has been given Versed, an antianxiety medication. He lies calmly on his side in a blue and white johnny with no pants. There is a heartbreaking vulnerability to people having hospital procedures. They may be CEOs or generals on the outside, but in here they are just patients, docile, hopeful, grateful.

The lights are dim and a stereo plays classical music. Khoruts makes conversation to gauge the sedative's effects. He's listening for a quieting of the voice, a slowing of words. "Do you have any pets?"

The room is quiet for a moment. ". . . pets."

"I think we're ready to go."

A nurse brings the bowl with the vials. I ask her if the red color of the caps on the vials signifies biohazard.

"No, just the brown color inside."

Unless one is watching closely, a fecal transplant looks very much like a colonoscopy. The first thing to appear on the video monitor is a careering fish-eye view of the exam room as the scope is pulled from its holder and carried over to the bed. If you are young enough to be unfamiliar with a colonoscope, I invite you to picture a bartender's soda gun: the long, flexible black tube, the controls mounted on a handheld head. Where the bartender has buttons for soda water and cola, Khoruts can choose between carbon dioxide, for inflating the colon so he can see it better, and saline, for rinsing away remnants of an "inadequate prep."

Khoruts works the control buttons with his left hand, torquing the tube with his right. I comment that it's like playing an accordion or a piano, both arms working independently at unrelated tasks. Khoruts, who plays piano in addition to colonoscope, prefers the analogy of the amputee's prosthesis. "Over time it becomes part of your body. Even though I don't have nerve endings there, I kind of know what's happening."

We're in now, heading north. The man's heartbeat is visible as a quiver in the colon wall. Khoruts maneuvers a crook. Shifting a patient's position can help unkink a sharp turn, so the nurse leans in hard, like a driver pushing a stall to the shoulder of the road.

Using a plunger on the control head, Khoruts releases a portion of the transplant material. Since the colon has been wiped clean beforehand with antibiotics, the unicellular arrivals won't have to battle a lot of natives. However many survived the antibiotic, the immigrants are sure to prevail. Within two weeks, Khoruts's research shows, the microbial profiles of donor and recipient colons are synced.

One more release, at the far end of the colon, and Khoruts retracts the scope.

A couple days later, Khoruts forwards an e-mail from the patient (with surname deleted). The pain and diarrhea that had kept him from going to work for a year were gone. "I had," he wrote, "a small solid bowel movement on Saturday evening." It may not be your idea of an exciting Saturday evening, but for Mr. F., it was tough to top.

• • •

THE FIRST FECAL transplant was performed in 1958, by a surgeon named Ben Eiseman. In the early days of antibiotics, patients frequently developed diarrhea from the massive kill-off of normal bacteria. Eiseman thought it might be helpful to restock the gut with someone else's normals. "Those were the days when if we had an idea," says Eiseman, ninety-three and living in Denver at the time I wrote him, "we simply tried it."

Rarely does medical science come up with a treatment so effective, inexpensive, and free of side effects. As I write this, Khoruts has done forty transplants to treat intractable *C. diff* infection, with a success rate of 93 percent. In a University of Alberta study published in 2012, 103 out of 124 fecal transplants resulted in immediate improvement. It's been fifty-five years since Eiseman first pushed the plunger, yet no U.S. insurance company formally recognizes the procedure.

Why? Has the "ick factor" hampered the procedure's acceptance? Partly, says Khoruts. "There is a natural revulsion. It just doesn't seem right." He thinks it has more to do with the process by which a new medical procedure goes from experimental to mainstream. A year after I visited, the major gastroenterology and infectious disease societies invited "a little band of fecal transplant practitioners" to put together a "best practice" paper outlining optimal procedures: a common first step toward establishing codes for billing for the procedure and making the case for insurance companies to cover it. As of mid-2012, there was no billing code or agreed-upon fee. Khoruts estimates the process will take one to two years more. In the meantime, he simply bills for a colonoscopy.

The extent to which health care bureaucracy stands in the way of better patient care is occasionally astounding. It took a year and a half for Khoruts's study on bacteriotherapy for recurrent *C. diff* infection to be approved by the University of Minnesota's Institutional Review Board (IRB)—which oversees the safety of study subjects—even though the board had no substantive criticisms or concerns. The morning I visited to see the transplant, Khoruts showed me an object I wasn't familiar with, a winged plastic bowl called a toilet hat* that fits over the rim of the bowl to catch the donor's produce. "That caused about two months of delay on the IRB protocol," he said. "They sent it back saying, 'Who's going to pay for the toilet hats?' They're fifty cents apiece."

Khoruts has also been working on a proposal for a study to evaluate fecal transplants for treating ulcerative colitis.† Inflammatory bowel diseases—irritable bowel syndrome, ulcerative colitis, Crohn's disease—are thought to be caused by an inappropriate immune response to normal bacteria; the colon gets caught in the cross fire. This time around, the IRB refused to approve the trial until the FDA had approved it. And that's just for the trial. Final FDA approval, the kind that makes the procedure available to anyone, is a costly process that can take upward of a decade.

And in the case of fecal transplants, there's no drug or medical device involved, and thus no pharmaceutical company or device maker with diverticula deep enough to fund the multiple rounds

---

* Or, less often, a nun's hat, because of the resemblance to the Flying Nun–style wimple. Catholic nurses and hospital patients have from time to time voiced their indignation, and the term has been mostly retired.

† Typing *colitis* reliably brings "Lucy in the Sky with Diamonds" into my head. In my favorite case of mistaken lyrics, someone heard "The girl with kaleidoscope eyes" as "The girl with colitis goes by."

of controlled clinical trials. If anything, drug companies might be inclined to fight the procedure's approval. Pharmaceutical companies make money by treating diseases, not by curing them. "There's billions of dollars at stake," says Khoruts. "I told Katerina, if this works, don't be surprised to find me at the bottom of the river."

We are sitting in Khoruts's office, in between colonoscopies. Above our heads, on a shelf, is a lurid plastic life-size model of a human rectum afflicted by every imaginable malady: hemorrhoid, fistula, ulcerative colitis, fecaliths. Metaphor for the U.S. health care system?

Khoruts smiles. "Bookend." A drug company was giving them away at Digestive Disease Week, an annual convention of gastroenterologists and drug reps, with the occasional person dressed as a stomach, handing out samples.

While the bureaucracy inches forward, fecal transplants for *C. diff* are quietly carried out in hospitals in thirty states. But that leaves twenty where patients have no access. Some have turned to what a researcher in one *Clinical Gastroenterology and Hepatology* paper called "self-administered home fecal transplantation." Though seven of seven *C. diff* sufferers were cured by self- or "family-administered" transplants using a drugstore enema kit, it doesn't always go well. One woman who recently e-mailed Khoruts for advice didn't follow directions. She put tap water in the blender, and the chlorine killed the bacteria. Another in-home transplant replaced one source of diarrhea with another: fecal parasites contracted from the donor. Rather than protecting patients, IRBs—with their delays and prodigious paperwork—can put them in harm's way.

Fecal bacteriotherapy will quickly become more streamlined. More sophisticated filtration will enable the separation of cellular material from ick. The bacteria can then be dosed with cryoprotectant—to prevent ice crystals from puncturing the cells—frozen, and shipped where it's needed, when it's needed. Khoruts's operation is already headed this way.

The Holy Grail would be a simple pill, along the lines of the lactobacillus suppositories used to cure recurrent yeast infection. Generally and unfortunately, aerobic strains that are easy to grow and keep alive in the oxygen environment of a lab are unlikely to be the beneficial ones. Though researchers don't know exactly which bacteria are the desirables, they do know they're likely to be anaerobic species that thrive only within the colon. You want the creatures that are dependent on a healthy you for their own survival, the ones whose evolutionary mission is aligned with your own—your microscopic partners in health.

I asked Khoruts what exactly is in the "probiotic" products seen in stores now. "Marketing," he replied. Microbiologist Gregor Reid, director of the Canadian Research & Development Centre for Probiotics, seconds the sentiment. With one exception, the bacteria (if they even exist) in probiotics are aerobic; culturing, processing, and shipping bacteria in an oxygen-free environment is complicated and costly. Ninety-five percent of these products, Reid told me, "have never been tested in a human and should not be called probiotic."

*I* PREDICT THAT ONE way or another, within a decade, everyone will know someone who's benefited from a dose of someone else's body products. I recently received an e-mail from a doctor in Texas, telling me the story of Lloyd Storr, a Lubbock physician who treated chronic ear infections via homemade "earwax transfusions": drops of donor earwax boiled up in glycerin. Earwax maintains an acid environment that discourages bacterial over-growth and possibly contains some antibacterial chemicals. Whatever it does, some people's works better than others'. Khoruts has been encouraging a friend of his, a periodontist, to try bacterial transplantation* as a treatment for gum disease.

If things go as they should, the bacteria hysteria so lucratively nurtured by the likes of Purell and Lysol will begin to subside. Thanks to the courageous blender-wielding pioneers of bacterial transplantation, fussiness and unfounded fear will be buffered by rational thinking and perhaps even a modicum of gratitude.

A tip of the toilet hat to you, Alexander Khoruts.

THE GREAT IRONY is that in the beginning, the gut was all there was. "We're basically a highly evolved earthworm sur-rounding the intestinal tract," Khoruts commented as we drove away from his clinic the last day I was there. Eventually, the food processor had to have a brain attached to help it look for food, and limbs to reach that food. That increased its size, so it needed a circulatory system to distribute the fuel that powered the limbs. And so on. Even now, the digestive tract has its own immune system and its own primitive brain, the so-called enteric nervous system. I recalled what Ton van Vliet had said at one point in our conversation: "People are surprised to learn: They are a big pipe with a little bit around it."

You are what you eat, but more than that, you are *how* you eat.

---

* Kissing is a less aggressive form of bacterial transplant. Studies of three different gingivitis-causing bacteria have documented migration from spouse to spouse. Periodontically speaking, an affair might be viewed as a form of bacteriotherapy.

Be thankful you're not a sea anemone, disgorging lunch through the same hole that dinner goes in. Be glad you're not a grazer or a cud chewer, spending your life stoking the furnace. Be thankful for digestive juices and enzymes, for villi, for fire and cooking, all the miracles that have made us what we are. Khoruts gave the example of the gorilla, a fellow ape held back by the energy demands of a less streamlined gut. Like the cow, the gorilla lives by fermenting vast quantities of crude vegetation. "He's processing leaves all day. Just sitting and chewing, and cooking inside. There's no room for great thoughts."

Those who know the human gut intimately see beauty, not only in its sophistication but in its inner landscapes and architecture. In a 1998 issue of the *New England Journal of Medicine*, two Spanish physicians published a pair of photographs: "the haustrations of the transverse colon" side by side with the arches of an upper-floor arcade in Gaudi's La Pedrera. Inspired, wanting to see my own internal Gaudi, I had my first colonoscopy without drugs.*

There is an unnameable feeling I've had maybe ten times in my life. It is a mix of wonder, privilege, humility. An awe that borders on fear. I've felt it in a field of snow on the outskirts of Fairbanks, Alaska, with the northern lights whipping overhead so seemingly close I dropped to my knees. I am walloped by it on dark nights in the mountains, looking up at the sparkling smear of our galaxy. Laying eyes on my own ileocecal valve, peering into my appendix from within, bearing witness to the magnificent complexity of the human body, I felt, let's be honest, mild to moderate cramping. But you understand what I'm getting at

---

* Not typically a big deal. Most Europeans get scoped with sedation-on-demand. You're set up with an IV ready to go, and need only say the word. Eighty percent never ask for the drugs.

here. Most of us pass our lives never once laying eyes on our organs, the most precious and amazing things we own. Until something goes wrong, we barely give them thought. This seems strange to me. How is it that we find Christina Aguilera more interesting than the inside of our own bodies? It is, of course, possible that I seem strange. You may be thinking, *Wow, that Mary Roach has her head up her ass.* To which I say: Only briefly, and with the utmost respect.

# *Acknowledgments*

This time around, I took a cue from the world of charitable giving. The categories below reflect the many levels of generosity and support that have made this book possible. If *Gulp* is interesting and fun, if it's accurate, enlightening, or compelling, that is due in overwhelming proportion to the contributions of these excellent human beings.

## PLATINUM UVULA CIRCLE

For giving up entire afternoons with no compensation and no guarantee of pleasing portrayal, for walking me through archives, for twisting arms, opening doors, laying out welcome mats, I bow down to:

Andrea Bainbridge, American Medical Association Historical Health Fraud and Alternative Medicine Collection

Ed dePeters, University of California, Davis

Anna Dhody and Evi Numen, Mütter Museum

Michael Jones, Virginia Commonwealth University

Alexander Khoruts, Matt Hamilton, and Mike Sadowsky, University of Minnesota

Alan Kligerman, Kligerman Regional Digestive Disease Center

Sue Langstaff, Applied Sensory

Michael Levitt and Julie Furne, Minneapolis VA Medical Center

George "Nick" Nichopoulos, personal physician to the late Elvis Presley

Megan and Rick Prelinger, Prelinger Library

Nancy Rawson, Pat Moeller, Amy McCarthy, and Theresa Kleinsorge, AFB International

"Rodriguez," Gene Parks, Ed Borla, and Paul Verke, Avenal State Prison and the California Department of Corrections and Rehabilitation

Stephen Secor, University of
Alabama
Erika Silletti, René de Wijk,
Andries van der Bilt, and

Ton van Vliet, Food Valley, the
Netherlands
Richard Tracy, Lee Lemenager,
and John Gray, University of
Nevada, Reno

## GOLDEN PYLORUS GUILD

For enduring repeated phone calls and protracted e-mail pestering with
no outward indication that the author had overstepped the bounds of
casual inquiry and was perched on the brink of actionable nuisance, I
salute:

Jianshe Chen
Phillip Clapham
Justin Crump
Evangelia Bellas
Thomas Lowry
David Metz
Jason Mihalopoulos

Gabriel Nirlungayuk
Adrianne Noe
Tom Rastrelli
Danielle Reed
Paul Rozin
Terrie Williams
Sera Young

## BRONZE BOLUS CLUB

For providing indispensable expertise on arcane topics, for sharing contacts, for inspiring and encouraging me and making me laugh, I thank:

Jaime Aranda-Michel
Dean Backer
Daniel Blackburn
Rabbi Zushe Blech
Laurie Bonneau
Andrea Chevalier
Patty Davis
Siobhan DeLancy
Erik "the Red" Denmark
Adam Drewnowski
Ben Eiseman
Holly Embree

Father Geoff Farrow
Richard Faulks
Steve Geiger
Roy Goodman
Farid Haddad
Susan Hogan
Al Hom
Tim Howard
Bruce Jayne
Mark Johnson
Mary Juno
Jason Karlawish

Ron Kean
Diane Kelly
Bruce Kraig
Christopher Lahr
Jennifer Long
Johan Lündstrom
Ray and Robert Madoff
The Notto
Kenneth Olson
Jon Prinz
Sarah Pullen

Gregor Reid
Janet Riley
Michael Sappol
Adam Savage
Markus Stieger
Jim Turner
Paul Wagner
Brian Wansink
Judge Colleen Weiland
William Whitehead

## SUSTAINING

For standing by me all these years and all these books, for their warmth, talent, patience, and friendship, a pixel and paper embrace to:

Jill Bialosky, Erin Lovett and Louise Brockett, Bill Rusin and Jeannie Luciano, and Stephen King and Drake McFeely of W. W. Norton, plus Mary Babcock, eagle-eyed copyeditor extraordinaire
Stephanie Gold

Jeff Greenwald
Jay Mandel and Lauren Whitney, of William Morris Endeavor
Lisa Margonelli
Anne Pigué
Ed and the rest of the wonderful Rachles family

# Bibliography

**INTRODUCTION**

Waslien, Carol, Doris Howes Calloway, and Sheldon Margen. "Human Intolerance to Bacteria as Food." *Nature* 221: 84–85 (January 4, 1969.)

**1 • NOSE JOB**

Drake, M. A., and G. V. Civille. "Flavor Lexicons." *Comprehensive Reviews in Food Science and Food Safety* 2: 33–40 (2003).

Hodgson, Robert T. "An Analysis of the Concordance among 13 U.S. Wine Competitions." *Journal of Wine Economics* 4 (1): 1–9 (Spring 2009).

Hui, Y. H. *Handbook of Fruit and Vegetable Flavors.* Hoboken: Wiley, 2010.

Mainland, Joel, and Noam Sobel. "The Sniff Is Part of the Olfactory Percept." *Chemical Senses* 31: 181–196 (2006).

Morrot, Gil, Frederic Brochet, and Denis Dubourdieu. "The Color of Odors." *Brain and Language* 79 (2): 309–320 (November 2001).

Mustacich, Suzanne. "Fighting Fake Bordeaux." *Wine Spectator,* November 8, 2011. www.winespectator.com/webfeature/show/id/45968.

Pickering, G. J. "Optimizing the Sensory Characteristics and Acceptance of Canned Cat Food: Use of a Human Taste Panel." *Journal of Animal Physiology and Animal Nutrition* 93 (1): 52–60 (February 2009).

Smith, Philip W., Owen W. Parks, and Daniel P. Schwartz. "Characterization of Male Goat Odors: 6-Trans Nonenal." *Journal of Dairy Science* 67 (4): 794–801 (April 1984).

**2 • I'LL HAVE THE PUTRESCINE**

Association of American Feed Control Officials. *Feed Ingredient Definitions,* Official Publication, 1992.

McCarrison, Robert. "A Good Diet and a Bad One: An Experimental Contrast." *British Medical Journal* 2 (3433): 730–732 (October 23, 1926).

Phillips, Tim. "Learn from the Past." *Petfood Industry* (October 2007). Pp. 14–17.

Wentworth, Kenneth L. "The Effect of a Native Mexican Diet on Learning and Reasoning in White Rats." *Journal of Comparative Psychology* 22 (2): 255–267 (October 1936).

## 3 · LIVER AND OPINIONS

*Apicius*. Book VIII: *Tetrapus (Quadrupeds)*.

Blake, Anthony A. "Flavour Perception and the Learning of Food Preferences." In *Flavor Perception*, edited by A. J. Taylor and D. D. Roberts. Hoboken: Wiley-Blackwell, 2004.

Blech, Zushe Yosef. "Like Mountains Hanging by a Hair." Kashrut.com. http://www.kashrut.com/articles/L_cysteine/ (accessed September 2012).

Bull, Sleeter. *Meat for the Table*. New York: McGraw-Hill, 1951.

Casteen, Marie L. "Ten Popular Specialty Meat Recipes." *Hotel Management*, August 1944. Pp. 26–28.

Cline, Jessie Alice. "The Variety Meats." *Practical Home Economics* 21: 57–58 (February 1943).

Davis, Clara. "Results of the Self-Selection of Diets by Young Children." *Canadian Medical Association Journal* 41 (3): 257–261 (September 1939).

Feeney, Robert E. *Polar Journeys: The Role of Food and Nutrition in Early Exploration*. Fairbanks: University of Alaska Press, 1997.

Guthe, Carl E., and Margaret Mead. "Manual for the Study of Food Habits: Report of the Committee on Food Habits." *Bulletin of the National Research Council*, No. 111 (1943.)

———. "The Problem of Changing Food Habits: Report of the Committee on Food Habits." *Bulletin of the National Research Council*, No. 108 (1943).

"Jackrabbit Should Be Used To Ease Meat Shortage." *Science News Letter*, July 24, 1943.

Kizlatis, Lilia, Carol Deibel, and A. J. Siedler. "Nutrient Content of Variety Meats." *Food Technology*, January 1964.

Kuhnlein, Harriet V., and Rula Soueida. "Use and Nutrient Composition of Traditional Baffin Inuit Foods." *Journal of Food Composition and Analysis* 5: 112–126 (1992).

Mead, Margaret. "Reaching the Last Woman down the Road." *Journal of Home Economics* 34: 710–713 (1942).

Mennella, J. A., and G. K. Beauchamp. "Maternal Diet Alters the Sensory Qualities of Human Milk and the Nursling's Behavior." *Pediatrics* 88 (4): 737–744 (1991).

Mennella, J. A., A. Johnson, and G. K. Beauchamp. "Garlic Ingestion by Pregnant Women Alters the Odor of Amniotic Fluid." *Chemical Senses* 20 (2): 207–209 (1995).

Rozin, Paul, et al. "Individual Differences in Disgust Sensitivity: Comparisons and Evaluations of Paper-and-Pencil versus Behavioral Measures." *Journal of Research in Personality* 33: 330–351 (1999).

———. "The Child's Conception of Food: Differentiation of Categories of Rejected Substances in the 16 Months to 5 Year Age Range. *Appetite* 7: 141–151 (1986).

Wansink, Brian. "Changing Eating Habits on the Home Front: Lost Lessons from World War II Research." *Journal of Public Policy and Marketing* 21 (1): 90–99 (Spring 2002).

Wansink, Brian, Steven T. Sonka, and Matthew M. Cheney. "A Cultural Hedonic Framework for Increasing the Consumption of Unfamiliar Foods: Soy Acceptance in Russia and Colombia." *Review of Agricultural Economics* 24 (2): 353–365 (September 23, 2002).

War Food Administration. *Food Conservation Education in the Elementary School Program* (pamphlet). Washington, D.C.: USDA, 1944.

## 4 · THE LONGEST MEAL

Barnett, L. Margaret. "Fletcherism: The Chew-Chew Fad of the Edwardian Era." In *Nutrition in Britain: Science, Scientists and Politics in the Twentieth Century*, edited by David Smith. London: Routledge, 1997.

———. "The Impact of 'Fletcherism' on the Food Policies of Herbert Hoover during World War I." *Bulletin of the History of Medicine* 66: 234–259 (June 1992).

Chittenden, Russell H. "The Nutrition of the Body: A Study in Economical Feeding." *Popular Science Monthly*, June 1903.

Dawson, Percy M. *A Biography of François Magendie.* Brooklyn: Albert T. Huntington, 1908.

"Eating Guano." *California Farmer and Journal of Useful Sciences* 11 (22) (July 1, 1859).

Fletcher, Horace. *The New Glutton or Epicure.* New York: Frederick A. Stokes, 1917.

Levine, Allen S., and Stephen E. Silvis. "Absorption of Whole Peanuts, Peanut Oil, and Peanut Butter." *New England Journal of Medicine* 303 (16): 917–918 (October 16, 1980).

## 5 · HARD TO STOMACH

Beaumont, William. *Experiments and Observations on the Gastric Juice, and the Physiology of Digestion.* Edinburgh: Maclachlan and Stewart, 1838.

Green, Alexa. "Working Ethics: William Beaumont, Alexis St. Martin, and Medical Research in Antebellum America." *Bulletin of the History of Medicine* 84 (2): 193–216 (Summer 2010).

Janowitz, Henry D. "Newly Discovered Letters concerning William Beaumont, Alexis St. Martin, and the American Fur Company." *Bulletin of the History of Medicine* 22 (6): 823–832 (November/December 2008).

Karlawish, Jason. *Open Wound: The Tragic Obsession of Dr. William Beaumont.* Ann Arbor: University of Michigan Press, 2011.

Leblond, Sylvio. "The Life and Times of Alexis St. Martin." *Canada Medical Association Journal* 88: 1205–1211 (June 15, 1963).

Myer, Jesse S. *Life and Letters of Dr. William Beaumont.* St. Louis: C. V. Mosby, 1912.

Roland, Charles G. "Alexis St. Martin and His Relationship with William Beaumont." *Annals of the Royal College of Physicians and Surgeons of Canada* 21 (1): 15–20 (January 1988).

## 6 · SPIT GETS A POLISH

"Breastfeeding Fatwa Sheikh Back at Egypt's Azhar." *Al Arabiya News,* May 18, 2009. http://www.alarabiya.net/articles/2009/05/18/73140.html.

Broder, J., et al. "Low Risk of Infection in Selected Human Bites Treated without Antibiotics." *American Journal of Emergency Medicine* 22 (1): 10–13 (January 2004).

Bull, J. J., Tim S. Jessup, and Marvin Whiteley. "Deathly Drool: Evolutionary and Ecological Basis of Septic Bacteria in Komodo Dragon Mouths." PloS One 5 (6): e11097 (June 21, 2010).

Chowdharay-Best, G. "Notes on the Healing Properties of Saliva." *Folklore* 75: 195–200 (1975).

Eastmond, C. J. "A Case of Acute Mercury Poisoning." *Postgraduate Medical Journal* 51: 428–430 (June 1975).

Fry, Brian, et al. "A Central Role for Venom in Predation by *Varanus komodoensis* (Komodo Dragon) and the Extinct Giant *Varanus (Megalania) priscus*. *Proceedings of the National Academy of Sciences* 106 (22): 8969–8974 (June 2, 2009).

Harper, Edward B. "Ritual Pollution as an Integrator of Caste and Religion." *Journal of East Asian Studies* 23: 151–197 (1964).

Hendley, J. Owen, Richard P. Wenzel, and Jack M. Gwaltney Jr. "Transmission of Rhinovirus Colds by Self-Inoculation." *New England Journal of Medicine* 288 (26): 1361–1364 (June 28, 1973).

Humphrey, Sue, and Russell T. Williamson. "A Review of Saliva: Normal Composition, Flow, and Function." *Journal of Prosthetic Dentistry* 85 (2): 162–169 (February 2001).

Hutson, J. M., et al. "Effect of Salivary Glands on Wound Contraction in Mice." *Nature* 279: 793–795 (June 28, 1979).

Jamjoon, Mohammed, and Saad Abedine. "Saudis Order 40 Lashes for Elderly Woman for Mingling." *CNN.com/world*, March 9, 2009. www.cnn.com/2009/WORLD/meast/03/09/saudi.arabia.lashes/index.html.

Kerr, Alexander Creighton. *The Physiological Regulation of Salivary Secretions in Man.* New York: Pergamon Press, 1961.

Lee, Henry. "On Mercurial Fumigation in the Treatment of Syphillis." *Medico-Chirurgical Transactions* 39: 339–346 (1856).

Lee, V. M., and R. W. A. Linden. "An Olfactory-Parotid Salivary Reflex in Humans?" *Experimental Physiology* 76: 347–355 (1991).

Mennen, U., and C. J. Howells. "Human Fight-Bite Injuries of the Hand: A Study of 100 Cases within 18 Months." *Journal of Hand Surgery* (British and European volume) 16 (4): 431–435 (November 1991).

Montgomery, Joel M., et al. "Aerobic Salivary Bacteria in Wild and Captive Komodo Dragons." *Journal of Wildlife Diseases* 38 (3): 545–551 (2002).

Nguyen, Sean, and David T. Wong. "Cultural, Behavioral, Social and Psychological Perceptions of Saliva: Relevance to Clinical Diagnostics." *CDA Journal* 34 (4): 317–322 (April 2006).

Oudhoff, Menno, et al. "Histatins Are the Major Wound-Closure Stimulating Factors in Human Saliva as Identified in a Cell Culture Assay." *FASEB Journal* 22: 3805–3812 (November 2008).

Patil, Pradnya D., Tanmay S. Panchabnai, and Sagar C. Galwankar. "Managing Human Bites." *Journal of Emergencies, Trauma, and Shock* 2 (3): 186–190 (September–December 2009).

Read, Bernard E. *Chinese Materia Medica: Animal Drugs, from the Pen Ts'ao Kang Mu by Li Shih-chen, A.D. 1597.* Taipei: Southern Materials Center, 1976.

Robinson, Nicholas. *A Treatise on the Virtues and Efficacy of a Crust of Bread: Eat Early in a Morning Fasting, to Which Are Added Some Particular Remarks concerning the Great Cures Accomplished by the Saliva or Fasting Spittle . . .* London: A. & C. Corbett, 1763.

Romão, Paula M. S., Adilia M. Alarcão, and Cesar A. N. Viana. "Human Saliva as a Cleaning Agent for Dirty Surfaces." *Studies in Conservation* 35: 153–155 (1990).

Rozin, Paul, and April E. Fallon. "A Perspective on Disgust." *Psychological Review* 94 (1): 23–41 (1987).

Silletti, Erika M. G. *When Emulsions Meet Saliva: A Physical-Chemical, Biochemical, and Sensory Study.* Thesis, Wageningen University, 2008.

## 7 · A BOLUS OF CHERRIES

Altkorn, Robert. "Fatal and Non-fatal Food Injuries among Children (Aged 0–14 Years)." *International Journal of Pediatric Otorhinolaryngology* 72 (7): 1041–1046 (July 2008).

Gliniecki, Andrew. "Elton John Wins Pounds 350,000 for Libel: Punitive Damages Awarded against 'Sunday Mirror' over False Claims about Diet." *Independent*, November 5, 1993.

Heath, M. R. "The Basic Mechanics of Mastication: Man's Adaptive Success." In *Feeding and the Texture of Food*, edited by J. F. V. Vincent. Cambridge, U.K.: Cambridge University Press, 2008.

*John v. MGN, Ltd.*, QB 586 (1997), 3 WLR 593 (1996), 2 All ER 35 (1996), EMLR 229 Court of Appeal, Civil Division (1996).

Mitchell, James E., et al. "Chewing and Spitting Out Food as a Clinical Feature of Bulimia." *Psychosomatics* 29: 81–84 (1988).

Prinz, Jon F., and René de Wijk. "The Role of Oral Processing in Flavour Perception." In *Flavor Perception*, edited by A. J. Taylor and D. D. Roberts. Hoboken: Wiley-Blackwell, 2004.

Seidel, James S., and Marianne Gausche-Hill. "Lychee-Flavored Gel Candies: A Potentially Lethal Snack for Infants and Children." *Archives of Pediatrics and Adolescent Medicine* 156 (11): 1120–1122 (November 2002).

Van der Bilt, Andries. "Assessment of Mastication with Implications for Oral Rehabilitation: A Review. *Journal of Oral Rehabilitation* 38: 754–780 (2011).

Wolf, Stewart. *Human Gastric Function: An Experimental Study of a Man and His Stomach.* Oxford, U.K.: Oxford University Press, 1947.

## 8 • BIG GULP

"A Shark Story of Great Merit." *New York Times*, December 4, 1896.

Bernard, Claude. *Leçons de Physiologie Expérimentale Appliquée a la Médecine, Faites au College de France.* Paris: Bailliere, 1855. Pp. 408–418.

Bondeson, J. "The Bosom Serpent." *Journal of the Royal Society of Medicine* 91: 442–447 (August 1998).

Dally, Ann. *Fantasy Surgery 1880–1930, with Special Reference to Sir William Arbuthnot Lane* (Clio Medica 38, Wellcome Institute Series in the History of Medicine). Amsterdam: Editions Rodopi B.V., 1996.

Dalton, J. C. "Experimental Investigations to Determine Whether the Garden Slug Can Live in the Human Stomach." *American Journal of Medical Sciences* 49 (98): 334–338 (April 1865).

Davis, Edward B. "A Whale of a Tale: Fundamentalist Fish Stories." *Perspectives on Science and Christian Faith* 43: 224–237 (1991).

Foster, Michael. *Lectures on the History of Physiology during the Sixteenth, Seventeenth, and Eighteenth Centuries.* Cambridge, U.K.: University Press, 1901.

Gambell, Ray, and Sidney G. Brown. "James Bartley—A Modern Jonah or a Joke?" *Investigations on Cetacea* 24: 325–337 (1993).

Hunter, John. "On the Digestion of the Stomach after Death." *Philosophical Transactions of the Royal Society* 62: 447–454 (1772).

Paget, Stephen. *Experiments on Animals.* London: James Nisbet, 1906.

Pavy, F. W. "On the Immunity Enjoyed by the Stomach from Being Digested by Its Own Secretion during Life." *Philosophical Transactions of the Royal Society* 153: 161–171 (1863).

Reese, D. Meredith. "Medical Curiosity: Alleged Living Reptile in the Human Stomach." *Boston Medical and Surgical Journal* 28 (18): 352–356 (June 7, 1908).

Slijper, E. J. *Whales.* New York: Basic Books, 1962. Pp. 284–293.

Spence, John. "Severe Affection of the Stomach, Ascribed to the Presence in It of an Animal of the Laerta Tribe." *Edinburgh Medical and Surgical Journal* 9: 315–318 (July 1813).

Stengel, Alfred. "Sensations Interpreted as Live Animals in the Stomach." *University of Pennsylvania Medical Bulletin* 16 (3): 86–89 (May 1903).

"Swallowed by a Whale." *New York Times,* November 22, 1896.

Warren, Joseph W. "Notes on the Digestion of 'Living' Tissues." *Boston Medical and Surgical Journal* 116 (11): 249–252 (March 17, 1887).

## 9 · DINNER'S REVENGE

Bland-Sutton, John. "The Psychology of Animals Swallowed Alive." In *On Faith and Science in Surgery.* London: William Heinemann, 1930.

Haddad, Farid S. "Ahmad ibn Aby al'Ash'ath (959 A.D.) Studied Gastric Physiology in a Live Lion." *Lebanese Medical Journal* 54 (4): 235 (2006).

Kozawa, Shuji, et al. "An Autopsy Case of Chemical Burns by Hydrochloric Acid." *Legal Medicine* 11: S535–S537 (2009).

Matshes, Evan W., Kirsten A. Taylor, and Valerie J. Rao. "Sulfuric Acid Injury." *American Journal of Forensic Medicine and Pathology* 29 (4): 340–345 (December 2008).

## 10 · STUFFED

Barnhart, Jay. S., and Roger E. Mittleman. "Unusual Deaths Associated with Polyphagia." *American Journal of Forensic Medicine and Pathology* 7 (1): 30–34 (1986).

Csendes, Atila, and Ana Maria Burgos. "Size, Volume, and Weight of the Stomach in Patients with Morbid Obesity Compared to Controls." *Obesity Surgery* 15 (8): 1133–1136 (September 2005).

Edwards, Gillian. "Case of Bulimia Nervosa Presenting with Acute Fatal Abdominal Distention." *Lancet* 325 (8432): 822–823 (April 6, 1985).

Glassman, Oscar. "Subcutaneous Rupture of the Stomach; Traumatic and Spontaneous." *Annals of Surgery* 89 (2): 247–263 (February 1929).

Key-Åberg, Algot. "Zur Lehre von der Spontanen Magenruptur." *Gerichtliche und Offfentliche Medicine* 3, 1: 42 (1891).

Lemmon, William T., and George W. Paschal Jr. "Rupture of the Stomach following Ingestion of Sodium Bicarbonate." *Annals of Surgery* 114 (6): 997–1003 (December 1941).

Levine, Marc S., et al. "Competitive Speed Eating: Truth and Consequences." *American Journal of Roentgenology* 189: 681–686 (2007).

Markowski, B. "Acute Dilatation of the Stomach." *British Medical Journal* 2 (4516): 128–130 (July 26, 1947).

Matikainen, Martti. "Spontaneous Rupture of the Stomach." *American Journal of Surgery* 138: 451–452 (September 1979).

Van Den Elzen, B. D., et al. "Impaired Drinking Capacity in Patients with Functional Dyspepsia: Intragastric Distribution and Distal Stomach Volume." *Neurogastroenterology and Motility* 19 (12): 968 –976 (December 2007).

## 11 • UP THEIRS

Agnew, Jeremy. "Some Anatomical and Physiological Aspects of Anal Sexual Practices." *Journal of Homosexuality* 12 (1): 75–96 (Fall 1985).

Cox, Daniel J., et al. "Additive Benefits of Laxative, Toilet Training, and Biofeedback Therapies in the Treatment of Pediatric Encopresis." *Journal of Pediatric Psychology* 21 (5): 659–670 (1996).

Garber, Harvey I., Robert J. Rubin, and Theodore E. Eisestat. "Foreign Bodies of the Rectum." *Journal of the Medical Society of New Jersey* 78 (13): 877–888 (December 1981).

Klauser, Andreas G., et al. "Behavioral Modification of Colonic Function: Can Constipation Be Learned?" *Digestive Diseases and Sciences* 35 (10): 1271–1275 (October 1990).

Knowlton, Brian, and Nicola Clark. "U.S. Adds Body Bombs to Concerns on Air Travel." *New York Times*, July 6, 2011.

Lancashire, M. J. R., et al. "Surgical Aspects of International Drug Smuggling." *British Medical Journal* 296: 1035–1037 (April 9, 1988).

Lowry, Thomas P., and Gregory R. Williams. "Brachioproctic Eroticism." *Journal of Sex Education and Therapy* 9 (1): 50–52 (1983).

Schaper, Andreas. "Surgical Treatment in Cocaine Body Packers and Body Pushers." *International Journal of Colorectal Disease* 22: 1531–1535 (2007).

Shafik, Ahmed, et al. "Functional Activity of the Rectum: A Conduit Organ or a Storage Organ or Both?" *World Journal of Gastroenterology* 12 (28): 4549–4552 (July 2006).

Simon, Gustav. "On the Artificial Dilatation of the Anus and Rectum for Exploration and for Operation." *Cincinnati Lancet and Observer* 14 (5): 326–334 (May 1873).

*State of Iowa v. Steven Landis*, Court of Appeals of Iowa, No. 1-500/10-1750 (2011).

Stephens, Peter J., and Mark L. Taff. "Rectal Impaction following Enema with Concrete Mix." *American Journal of Forensic Medicine and Pathology* 8 (2): 179–182 (1987).

*United States v. Delaney Abi Odofin*, 929 F.2d at 60.

*United States v. Montoya de Hernandez*, 473 U.S. 531 (1985).

Voderholzer, W. A., et al. "Paradoxical Sphincter Contraction Is Rarely Indicative of Anismus." *Gut* 41: 258–262 (1997).

Wetli, Charles V., Arundathi Rao, and Valerie Rao. "Fatal Heroin Body Packing." *American Journal of Forensic Medicine and Pathology* 18 (3): 312–318 (September 1997).

Yegane, Rooh-Allah, et al. "Surgical Approach to Body Packing." *Diseases of the Colon and Rectum* 52 (1): 97–103 (2009).

## 12 · INFLAMMABLE YOU

Avgerinos, A., et al. "Bowel Preparation and the Risk of Explosion during Colonoscopic Polypectomy." *Gut* 25: 361–364 (1984).

Bigard, Marc-Andre, Pierre Gaucher, and Claude Lassalle. "Fatal Colonic Explosion during Colonoscopic Polypectomy." *Gastroenterology* 77: 1307–1310 (1979).

Manner, Hendrik, et al. "Colon Explosion during Argon Plasma Coagulation." *Gastrointestinal Endoscopy* 67 (7): 1123–1127 (June 2008).

"Manure Pit Hazards." *Farm Safety & Health Digest* 3 (4, part 3).

McNaught, James. "A Case of Dilatation of the Stomach Accompanied by the Eructation of Inflammable Gas." *British Medical Journal* 1 (1522): 470–472 (March 1, 1890).

**13 · DEAD MAN'S BLOAT**

Beazell, J. M., and A. C. Ivy. "The Quality of Colonic Flatus Excreted by the 'Normal' Individual." *American Journal of Digestive Diseases* 8 (4): 128–132 (1941).

Furne, J. K., and M. D. Levitt. "Factors Influencing Frequency of Flatus Emission by Healthy Subjects." *Digestive Diseases and Sciences* 41 (8): 1631–1635 (August 1996).

Greenwood, Arin. "Taste-Testing Nutraloaf." *Slate*, June 24, 2008.

Kirk, Esben. "The Quantity and Composition of Human Colonic Flatus." *Gastroenterology* 12 (5): 782–794 (May 1949).

Levitt, Michael D., et al. "Studies of a Flatulent Patient." *New England Journal of Medicine* 295: 260–262 (July 29, 1976).

Magendie, F. "Note sur les gaz inestinaux de l'homme sain." *Annales de Chimie et de Physique* 2: 292 (1816).

Suarez, Fabrizis L., and Michael D. Levitt. "An Understanding of Excessive Intestinal Gas." *Current Gastroenterology Reports* 2: 413–419 (2000).

**14 · SMELLING A RAT**

Burkitt, D. F., A. R. P. Walker, and N. S. Painter. "Effect of Dietary Fibre on Stools and Transit-Times, and Its Role in the Causation of Disease." *Lancet* 300 (7792): 1408–1411 (December 30, 1972).

Donaldson, Arthur. "Relation of Constipation to Intestinal Intoxication." *Journal of the American Medical Association* 78 (12): 882–888 (March 25, 1922).

"Fatalities Attributed to Entering Manure Waste Pits—Minnesota, 1992." *MMWR Weekly* 42 (17): 325–329 (May 7, 1993).

Goode, Erica. "Chemical Suicides, Popular in Japan, Are Increasing in the U.S." *New York Times*, June 18, 2011.

Levitt, Michael D., and William C. Duane. "Floating Stools: Flatus versus Fat." *New England Journal of Medicine* 286 (18): 973–975 (May 4, 1972).

Kellogg, J. H. *The Itinerary of a Breakfast.* Battle Creek, Mich.: Modern Medicine Publishing, 1918.

Knight, Laura D., and S. Erin Presnell. "Death by Sewer Gas: Case Report of a Double Fatality and Review of the Literature." *American Journal of Forensic Medicine and Pathology* 26 (2): 181–185 (June 2005).

Moore, J. G., B. K. Krotoszynski, and H. J. O'Neill. "Fecal Odorgrams: A Method for Partial Reconstruction of Ancient and Modern Diets." *Digestive Diseases and Sciences* 29 (10): 907–912 (October 1984).

Oesterhelweg, L., and K. Puschel. "'Death May Come on Like a Stroke of Lightening . . .': Phenomenological and Morphological Aspects of Fatalities Caused by Manure Gas." *International Journal of Legal Medicine* 122: 101–107 (2008).

Ohge, Hiroki, et al. "Effectiveness of Devices Purported to Reduce Flatus Odor." *American Journal of Gastroenterology* 100 (2): 397–400 (February 2005).

Olson, K. R. "The Therapeutic Potential of Hydrogen Sulfide: Separating Hype from Hope." *American Journal of Physiology: Regulatory, Integrative and Comparative Physiology* 301 (2): R297–R312 (August 2011).

Osbern, L. N., and Crapo, R. O. "Dung Lung: A Report of Toxic Exposure to Liquid Manure." *Annals of Internal Medicine* 95 (3): 312–314 (1981).

Simons, C. C., et al. "Bowel Movement and Constipation Frequencies and the Risk of Colorectal Cancer among Men in the Netherlands Cohort Study on Diet and Cancer." *American Journal of Epidemiology* 172 (12): 1404–1414 (December 15, 2010).

Suarez, Fabrizis L., and Michael D. Levitt. "An Understanding of Excessive Intestinal Gas." *Current Gastroenterology Reports* 2: 413–419 (2000).

Suarez, F. L., J. Springfield, and M. D. Levitt. "Identification of Gases Responsible for the Odour of Human Flatus and Evaluation of a Device Purported to Reduce This Odor." *Gut* 43 (1): 100–104 (July 1998).

Walker, A. R. P. "Diet, Bowel Motility, Faeces Composition, and Colonic Cancer." *South African Medical Journal* 45 (14): 377–379 (April 3, 1971).

Whorton, James C. *Inner Hygiene: Constipation and the Pursuit of Health in Modern Society.* New York: Oxford University Press, 2000. Pp. 11–17.

Wild, P., et al. "Mortality among Paris Sewer Workers." *Occupational and Environmental Medicine* 63 (3): 168–172 (March 2006).


**15** · EATING BACKWARD

Armstrong, B. K., and A. Softly. "Prevention of Coprophagy in the Rat: A New Method." *British Journal of Nutrition* 20 (3): 595–598 (September 1966).

Barnes, Richard H. "Nutritional Implications of Coprophagy." *Nutrition Reviews* 20 (10): 289–291 (October 1962).

Barnes, Richard H., et al. "Prevention of Coprophagy in the Rat." *Journal of Nutrition* 63: 489–498 (1957).

Bertolani, Paco, and Jill Pruetz. "Seed-Reingestion in Savannah Chimpanzees (*Pan troglodytes verus*) at Fongoli, Senegal." *International Journal of Primatology* 32 (5): 1123–1132 (2011).

Bliss, D. W. *Feeding per Rectum.* Washington, D.C.: D. W. Bliss, M.D., 1882.

Bouchard, Charles. *Lectures on Autointoxication in Disease.* Philadelphia: F. A. Davis, 1898. Lecture 9, pp. 94–96.

Bugle, Charles, and H. B. Rubin. "Effects of a Nutritional Supplement on Corprophagia: A Study of Three Cases." *Research in Developmental Disabilities* 14: 445–446 (1993).

Dawson, W. W. "Bowel Exploration, Simon's Plan, Experiments upon the Cadaver." *Cincinnati Lancet and Clinic* 53 (14): 221–226 (1885).

Furst, Peter T., and Michael D. Coe. "Ritual Enemas." *Natural History*, March 1977. Pp. 88–91.

Herter, Christian Archibald. *The Common Bacterial Infections of the Digestive Tract and the Autointoxications Arising from Them.* New York: Macmillan, 1907.

Jones, L. E., and W. E. Norris. "Rectal Burn Induced by Hot Coffee Enema." *Endoscopy* 42: E26 (2010).

Kellogg, J. H. *The Itinerary of a Breakfast.* Battle Creek, Mich.: Modern Medicine, 1918.

Lane, Sir William Arbuthnot. "The Results of the Operative Treatment of Chronic Constipation." *British Medical Journal* 1: 126–130 (January 18, 1908).

Madding, Gordon F., Paul A. Kennedy, and R. Thomas McLaughlin.

"Clinical Use of Anti-Peristaltic Bowel Segments." *Annals of Surgery* 161 (4): 601–604 (April 1965).

Mutch, N., and J. H. Ryffel. "The Metabolic Utility of Rectal Feeding." *British Medical Journal* 1 (2716): 111–112 (January 18, 1913).

Onishi, Norimitsu. "From Dung to Coffee Brew with No Aftertaste." *New York Times* (Asia Pacific), April 17, 2010.

Rabino, A. "Storia della medicina: parabola di un prezioso alleato della vecchia medicina." *Minerva Medica* 43:459–466 (February 3, 1972).

Sammet, Kai. "Avoiding Violence by Technologies? Rectal Feeding in German Psychiatry." *History of Psychiatry* 17: 259–278 (2006).

Short, A. R., and H. W. Bywaters. "Amino-Acids and Sugars in Rectal Feeding." *British Medical Journal* 1 (2739): 1361–1367 (June 28, 1913).

## 16 · I'M ALL STOPPED UP

Battey, Robert. "A Safe and Ready Method of Treating Intestinal Obstruction." *Practitioner* 13: 441 (July–December 1874).

Black, Patrick. "Clinical Lecture on Obstinate Constipation and Obstruction of the Bowels." *British Medical Journal*, January 28, 1871. Pp. 83–84.

Corman, Marvin. "Classic Articles in Colon and Rectal Surgery: Sir William Arbuthnot Lane, 1856–1943." *Diseases of the Colon and Rectum* 28 (10): 751–757 (October 1985).

Dawson, W. W. "Bowel Exploration, Simon's Plan—Experiments upon the Cadaver—Introduction of the Hand . . ." *Cincinnati Lancet and Clinic* 14 (53): 221–226 (1885).

Formad, Henry F. "A Case of Giant Growth of the Colon, Causing Coprostasis, or Habitual Constipation." *Transactions of the College of Philadelphia* 14 (Series 3): 112–125 (1892).

Geib, D., and J. D. Jones. "Unprecedented Case of Constipation." *Journal of the American Medical Association* 38: 1304–1305 (May 17, 1902).

Klauser, A. G., et al. "Abdominal Wall Massage: Effect on Colonic Function in Healthy Volunteers and in Patients with Chronic Constipation." *Zeitschrift fur Gastroenterologie* 30 (4): 247–251 (April 1992).

Kollef, Marin H., and David T. Schachter. "Acute Pulmonary Embolism Triggered by the Act of Defecation." *Chest* 99 (2): 373–376 (1991).

Lahr, Chris. *Why Can't I Go?* Charlston, S.C.: Sunburst Press, 2004.

McGuire, Johnson, et al. "Bed Pan Deaths." *American Practitioner and Digest of Treatment* 1: 23–28 (1950).

Nichopoulos, George (with Rose Clayton Phillips). *The King and Dr. Nick: What Really Happened to Elvis and Me.* Nashville, Tenn.: Thomas Nelson, 2009.

Sikirov, B. A. "Cardio-vascular Events at Defecation: Are They Unavoidable?" *Medical Hypotheses* 32: 231–233 (1990).

Sydenham, Thomas. *The Works of Thomas Sydenham*, vol. 1. London: Sydenham Society, 1843.

Thompson, Charles C., and James P. Cole. *The Death of Elvis.* New York: Bantam Doubleday, 1991.

Wangensteen, Owen H. "Historical Aspects of the Management of Acute Intestinal Obstruction." *Surgery* 65 (2): 363–383 (1969).

Wide, Gustaf A. *Hand-Book of Medical and Orthopedic Gymnastics.* New York: Funk and Wagnalls, 1909.

**17 • THE ICK FACTOR**

Khoruts, A., et al. "Changes in the Composition of the Human Fecal Microbiome after Bacteriotherapy for Recurrent *Clostridium difficile–*Associated Diarrhea." *Journal of Clinical Gastroenterology* 44 (5): 354–360 (May/June 2010).

Martinez, Anna Paula, and Gisele Regina de Azevedo. "The Bristol Stool Form Scale: Its Translation to Portuguese, Cultural Adaptation, and Validation." *Revista Latino-Americana de Enfermagem* 20 (3): 583–589 (May/June 2012).

Offenbacher, S., B. Olsvik, and A. Tonder. "The Similarity of Periodontal Microorganisms between Husband and Wife Cohabitants. Association or Transmission?" *Journal of Periodontology* 56 (6): 317–323 (June 1985).

Parker-Pope, Tara. "Probiotics: Looking underneath the Yogurt Label." *New York Times* (Science Times column "Well"), September 28, 2009.

Silverman, Michael S., Ian Davis, and Dylan R. Pillai. "Success of Self-Administered Home Fecal Transplantation for Chronic *Clostridium Difficile* Infection." *Clinical Gastroenterology and Hepatology* 8 (5): 471–473 (May 2010).

Steenbergen, T. J., et al. "Transmission of *Porphyromonas gingivalis* between Spouses." *Journal of Clinical Periodontology* 20 (5): 340–345 (May 1993).

Terruzzi, Vittorio, et al. "Unsedated Colonoscopy: A Neverending Story." *World Journal of Gastrointestinal Endoscopy* 4 (4): 137–141 (April 16, 2012).

Willing, Benjamin P., and Janet K. Jansson. "The Gut Microbiota: Ecology and Function." In *The Fecal Bacteria*, edited by M. J. Sadowsky and R. L. Whitman. Washington, D.C.: American Society for Microbiology, 2011.